Springer Theses

Recognizing Outstanding Ph.D. Research

Aims and Scope

The series "Springer Theses" brings together a selection of the very best Ph.D. theses from around the world and across the physical sciences. Nominated and endorsed by two recognized specialists, each published volume has been selected for its scientific excellence and the high impact of its contents for the pertinent field of research. For greater accessibility to non-specialists, the published versions include an extended introduction, as well as a foreword by the student's supervisor explaining the special relevance of the work for the field. As a whole, the series will provide a valuable resource both for newcomers to the research fields described, and for other scientists seeking detailed background information on special questions. Finally, it provides an accredited documentation of the valuable contributions made by today's younger generation of scientists.

Theses are accepted into the series by invited nomination only and must fulfill all of the following criteria

- They must be written in good English.
- The topic should fall within the confines of Chemistry, Physics, Earth Sciences, Engineering and related interdisciplinary fields such as Materials, Nanoscience, Chemical Engineering, Complex Systems and Biophysics.
- The work reported in the thesis must represent a significant scientific advance.
- If the thesis includes previously published material, permission to reproduce this must be gained from the respective copyright holder.
- They must have been examined and passed during the 12 months prior to nomination.
- Each thesis should include a foreword by the supervisor outlining the significance of its content.
- The theses should have a clearly defined structure including an introduction accessible to scientists not expert in that particular field.

Indexed by zbMATH.

More information about this series at http://www.springer.com/series/8790

Vito Paolo Pastore

Estimating Functional Connectivity and Topology in Large-Scale Neuronal Assemblies

Statistical and Computational Methods

Doctoral Thesis accepted by
The University of Genova, Italy

Springer

Author
Dr. Vito Paolo Pastore
Department of Informatics, Bioengineering,
Robotics and System Engineering (DIBRIS)
University of Genova
Genova, Italy

Supervisor
Prof. Sergio Martinoia
Department of Informatics
Bioengineering, Robotics
and System Engineering (DIBRIS)
University of Genova
Genova, Italy

Institute of Biophysics
National Research Council (CNR)
Genova, Italy

ISSN 2190-5053 ISSN 2190-5061 (electronic)
Springer Theses
ISBN 978-3-030-59041-3 ISBN 978-3-030-59042-0 (eBook)
https://doi.org/10.1007/978-3-030-59042-0

This Springer imprint is published by the registered company Springer Nature Switzerland AG
The registered company address is: Gewerbestrasse 11, 6330 Cham, Switzerland

Parvum addas parvo,
magnus acervum erit

Supervisor's Foreword

It is a great pleasure to introduce the thesis work of Dr. Vito Paolo Pastore accepted for publication within Springer Theses and awarded with a prize for an outstanding original work. Vito Paolo Pastore performed his Master thesis under my supervision and then joined my research group on Neuro-engineering (Department of Informatics, Bioengineering, Robotics and Systems Engineering) for his Ph.D. program on Bioengineering and Robotics at the University of Genova.

He started in November 2014 and completed his studies with a successful thesis defense on February 13, 2018. The thesis reports an innovative study dealing with high-density large-scale microelectrode arrays (HDMEAs) and with the specific design and implementation of new algorithms for the characterization of the electrophysiological behavior of neuronal cultures coupled to such HDMEAs. Part of the presented results has been awarded by the National Bioengineering Group with the "Award for young scientists" in 2018, and most of them have been published in Plos Computational Biology on August 2018. Other results have been preliminary published on Neuroinformatics on January 2018.

In vitro neuronal networks for studying neuronal physiological and pathological functions are well-established experimental models in neuroscience since decades. They are widely used for molecular, cellular and networks studies, mainly for their possibility of manipulation, control and long-term culturing. They have recently experienced a new attention with the advent of human models thanks to the introduction of induced pluripotent stem cells (iPSCs). In the last decade, there was an increasing attention in the development of cultured networks from human iPSCs by recognizing the advantages, from the patho-physiological point of view of such preparations, to model neuronal diseases for translational research. Specific interest in functional connectivity studies related to the introduction of the concept of connectome has experiences as well a new and renowned interest. Notwithstanding the increasing attention of human in vitro neuronal models coupled to high-density MEAs, most of the electrophysiological studies are still lacking of new and ad hoc algorithms for reliable functional-effective connectivity estimation.

In his thesis work, Vito Paolo Pastore greatly contributed to the development of new experimental algorithms and their innovative implementation specifically tailored for in vitro models coupled to HDMEAs. These new computational tools were systematically validated first with computational model systems and then with experimental model systems. His work included new analysis tools (design and implementation) as well as optimization of computational models and experimental validations. Vito Paolo Pastore demonstrated the possibility of estimating both excitatory and inhibitory connection in large scale in vitro neuronal networks coupled to HDMEAs.

This validated analysis method and approach opens new possibilities in the network electrophysiology field, allowing estimating functional-effective connectivity maps in vitro neuronal systems supporting characterization for the many neural diseases in which the balance excitation–inhibition is compromised. Moreover, it constitutes a step forward in the field of neuro-engineering with specific relevance for the support to the analysis of human-based in vitro experimental model systems under the perspective of personalized and precision medicine.

Genoa, Italy
June 2020

Prof. Sergio Martinoia

Parts of this thesis have been published in the following documents:

Journal Publications

Pastore, V. P., Massobrio, P., Godjoski, A. & Martinoia, S. Excitatory-inhibitory links and network topology in large scale neuronal assemblies from multi-electrode recordings. PLoS Comput Biol. 2018;14(8): e1006381. Published 2018 Aug 27. https://doi.org/10.1371/journal.pcbi.100638

Pastore, V. P., Godjoski, A., Martinoia, S. & Massobrio, P. SPICODYN: A Toolbox for the analysis of neuronal network dynamics and connectivity from multi-site spike signal recordings. Neuroinformatics, https://doi.org/10.1007/s12021-017-9343-z (2017).

Pastore, V. P., Poli, D., Godjoski, A., Martinoia, S. & Massobrio, P. ToolConnect: a functional connectivity toolbox for in vitro networks. Frontiers in Neuroinformatics 10, https://doi.org/10.3389/fninf.2016.00013 (2016).

Poli, D., Pastore, V. P., Martinoia, S. & Massobrio, P. From functional to structural connectivity using partial correlation in neuronal assemblies. Journal of Neural Engineering 13, 026023 (2016).

Poli, D., Pastore, V. P. & Massobrio, P. Functional connectivity in in vitro neuronal assemblies. Frontiers in Neural Circuits 9, https://doi.org/10.3389/fncir.2015.00057 (2015).

Contents

1 Introduction 1
1.1 Bridging Structure and Function of Neural Circuits 1
1.2 Different Types of Connectivity to Describe Neuronal Assemblies 2
1.3 The Relevance of the In Vitro Model 3
1.4 Micro Electrode Arrays 3
1.5 How to Infer Functional Connectivity in In Vitro Neural Networks 4
1.6 Functional Connectivity Analysis Requires Ad-Hoc Software and Algorithms Implementation 6
1.7 Goal of the Thesis 7
1.8 Structure of the Thesis 8
References 8

2 Materials and Methods 11
2.1 Correlation-Based Connectivity Methods 11
2.1.1 Cross-Correlation 11
2.1.2 Cross-Correlation Histogram 12
2.1.3 Filtered and Normalized Cross-Correlation Histogram (FNCCH) 13
2.1.4 Partial Correlation 15
2.2 Information-Theory Based Connectivity Methods 17
2.2.1 Transfer Entropy 17
2.2.2 Delayed Transfer Entropy 18
2.2.3 Delayed High Order Transfer Entropy 19
2.2.4 Joint Entropy 19
2.3 Thresholding Approaches 20
2.3.1 Hard Threshold 21
2.3.2 Shuffling Approach 21

2.4 Metrics to Evaluate the Connectivity Methods' Performances 24
2.4.1 Receiver Operating Characteristic (ROC) Curve 24
2.4.2 Matthews Correlation Coefficient (MCC) Curve 25
2.5 Graph Theory 25
2.6 BIOCAM and Active Pixel Sensor (APS) Array 29
References 30

3 Results 33
3.1 Transfer Entropy 34
3.1.1 Validation of the TE by Means of in Silico Neural Networks 34
3.1.2 DTE Application to Neural Networks Coupled to the MEA-4k 36
3.2 Joint Entropy 36
3.2.1 Shuffling Thresholding Procedure Validation on in Silico Networks 36
3.2.2 Hard Threshold Validation on in Silico Neural Networks 38
3.2.3 JE Application to Cortex Network Coupled to the HD-MEA 41
3.3 FNCCH Results 43
3.3.1 Validation of the FNCCH by Means of in Silico Neural Networks 43
3.3.2 Functional Connectivity and Emergent Network Topologies in In Vitro Large-Scale Neural Networks 45
3.4 FNCCH Results: Discussion and Observations 50
3.4.1 Identification of Inhibition 51
3.4.2 The Emergence of a Scale Free and Small-World Topology 52
3.4.3 Comparison with a Transfer Entropy Based Algorithm . . . 53
3.4.4 Partial Correlation 54
3.4.5 Validation of the PC by Means of in Silico Neural Networks 54
3.4.6 PC Application to Experimental Data 56
3.4.7 Connectivity During Development 56
3.5 Algorithm Optimization 59
3.5.1 FNCCH: Algorithm Optimization 60
3.5.2 Transfer Entropy Algorithm Optimization 61
3.5.3 Joint Entropy Algorithm Optimization 62
3.6 Software Development for Functional Connectivity Analysis 63
3.6.1 Main Graphical Interface 63
3.6.2 Example of ToolConnect's Application to Experimental Data 64
3.6.3 Multi-threading Implementation 68

3.6.4 Input Data Processing and Conversion Procedures 69
3.6.5 Spiking and Bursting Features Analysis 71
3.6.6 Spike Detection 71
3.6.7 Burst Detection 72
3.6.8 Spiking and Bursting Statistics 73
3.6.9 Comparison with Other Software 75
3.6.10 Use of SpiCoDyn to Process Long-Lasting Recordings on MEA-4k 76
3.6.11 Dynamic Characterization 77
3.6.12 Connectivity Characterization 77
References 77

Conclusion 81

Appendix 85

References 87

Acronyms

APS	Active Pixel Sensor
CC	Cluster Coefficient
CCH	Cross-Correlation Histogram
DHOTE	Delayed High-Order Transfer Entropy
DTE	Delayed Transfer Entropy
FN	False Negative
FNCCH	*Filtered Normalized Cross-Correlation Histogram*
FP	False Positive
FPR	False Positive Rate
id	in-degree
JE	Joint Entropy
MCC	Matthews Correlation Coefficient Curve
MEA	Micro-Electrodes Array
NCCH	Normalized Cross-Correlation Histogram
od	out-degree
PC	Partial Correlation
PL	Path Length
PSTH	Post-Stimulus Time Histogram
RCC	Rich-Club Coefficient
ROC	Receiver Operating Characteristic Curve
SWI	Small World Index
SWM	Synaptic Weight Matrix
TCM	Thresholded Connectivity Matrix
td	total degree
TE	Transfer Entropy
TN	True Negative
TP	True Positive
TPR	True Positive Rate

Chapter 1
Introduction

1.1 Bridging Structure and Function of Neural Circuits

One of the most fundamental features of a neural circuit is its connectivity since the single neuron activity is not due only to its intrinsic properties but especially to the direct or indirect influence of other neurons [1]. As recently reviewed by Yuste [2], thanks to technology advancements, the *era* of the "neuron doctrine" has faded: neuronal assemblies can be considered the physiological units of the brain which generate and sustain the functional properties as well as the dynamical states of the entire system. Nervous systems are complex networks *par excellence*, capable of elaborating and integrating information from multiple external and internal sources in real time. It has been proposed that such information is managed by means of neural networks complying with two competing demands, which might also be considered as fundamental organizational principles: *functional segregation* and *functional integration*, enabling both the rapid extraction of information and the generation of coherent brain states [3]. As confirmed by recent studies reporting structural analyses of brain networks carried out on datasets describing the cerebral cortex of mammalian animal models (e.g. rat, cat, monkey), cortical areas were found to be neither completely connected with each other nor randomly linked; instead, their interconnections show a specific and intricate organization [4]. It is fundamental to elaborate and adopt research strategies aimed at a comprehensive structural description of such interconnections as well as the networks' elements forming the human connectome [5]. The connectome will significantly increase our understanding of how functional brain states emerge from their underlying structural substrate, and will provide new mechanistic insights into how brain function is affected if this structural substrate is disrupted.

V. P. Pastore, *Estimating Functional Connectivity and Topology in Large-Scale Neuronal Assemblies*, Springer Theses, https://doi.org/10.1007/978-3-030-59042-0_1

1.2 Different Types of Connectivity to Describe Neuronal Assemblies

The types of connectivity used to describe the interactions of neuronal networks are: structural, functional and effective.

Structural connectivity (Fig. 1.1a): the structural or anatomical connectivity indicates the physical interactions (i.e., a chemical or electrical synapses) that link network's neurons at a given time [6]. Therefore, we can determine which neural units can directly interact with each other. The structural connectivity ranges over multiple spatial scales, since we can detect morphological connections both in local microcircuits and in long-range interactions that link different sub-networks. In a short time scale (about., less than one minute), such morphological connections mediated by dendritic spines are static, while in a longer time scale, they are dynamic, since physiological mechanisms of learning, plasticity and development can shape the morphological circuits [7].

Functional connectivity (Fig. 1.1b): functional connection indicates the correlation between time series of spikes coming from different neurons. It measures statistical interdependence without considering any causal effects; it is time-dependent and "model-free". Therefore, two neurons are functionally linked, if we can predict the activity of one of the two neurons on the basis of the activity of the other neuron. Functional properties of single neurons are strongly driven by their anatomical connections, dendritic arborizations and synaptic distributions [6]. Moreover, functional

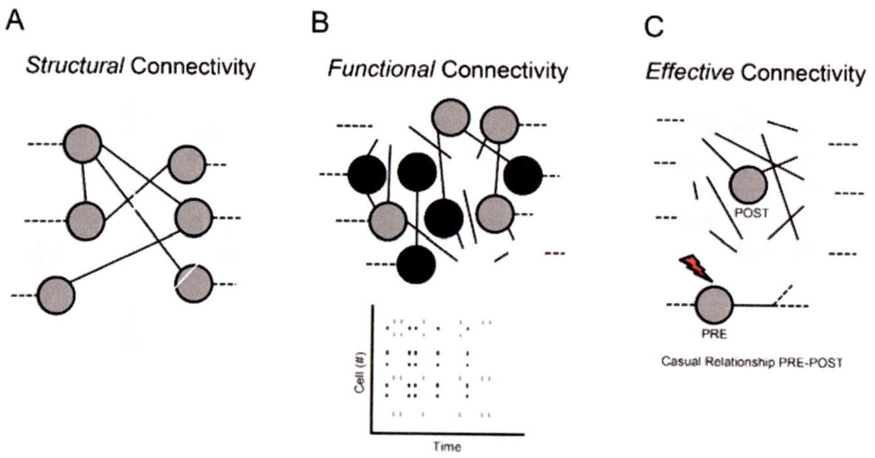

Fig. 1.1 Classification of the neural network connections. **a** Structural connectivity involving the physical connections between neurons. **b** Functional connectivity, involving functional connections and activity synchrony and correlation between neurons, without any causal model. **c** Effective connectivity. It is derived from functional connectivity introducing a causal model. Reproduced from [8]

interactions can contribute to the shaping of the underlying anatomical substrate through activity-dependent synaptic modifications.

Effective connectivity (Fig. 1.1c): effective connectivity indicates the presence of a connection when a neuron on the network directly affects another neuron through a causal relationship between the activities of the two neurons. In other words, "effective" means any observable interactions between neurons that alters their firing activity; so it is not "model-free" like functional connectivity, but can require the specification of a causal model including structural parameters.

1.3 The Relevance of the In Vitro Model

The final goal of connectivity analysis is the reconstruction of the human connectome, thus, the application of statistical measures to the in vivo model in both physiological and pathological states. Since the system under study (i.e. brain areas, cell assemblies) is highly complex and not controllable, to achieve the purpose described above, it is useful to adopt a reductionist approach. A possible strategy to reduce such a complexity makes use of in vitro experimental models coupled to Micro-Electrode Arrays (MEAs). Nowadays, dissociated neuronal cultures coupled to MEAs are fairly used to better understand the complexity of brain networks. One of the main advantage to use dissociated neuronal assemblies is the possibility to manipulate and control their connectivity: in other words, it is feasible to drive the connectivity of a network and to study how such a topological configuration can shape the emergent dynamics. Examples of engineered networks started in 1975 with the pioneering work of Letourneau [9] who investigated the role of different adhesion substrates for promoting the initiation, elongation and branching of the axons. The possibility to use the in vitro model, that is a valuable but at the same time reduced and simplified experimental model can be considered a great breakthrough in understanding the functional properties of neuronal networks.

1.4 Micro Electrode Arrays

MEAs are a powerful tool for simultaneously monitoring and acquiring the electrophysiological activity of neural preparations at many sites (Fig. 1.2a). The electrodes embedded in such devices can record electrophysiological activity in a non-invasive way (i.e., extracellularly) and therefore, under proper maintenance conditions, can allow long-term recordings (i.e., from hours up to months) [10]. Commercial available MEAs usually provide 60–120 electrodes with 100–500 μm inter-electrode spacing (Fig. 1.2b). However, recent advances in multichannel recording techniques have made possible to observe the activities of thousands of neurons simultaneously, and made routine the acquisition of massive amounts of empirical data or

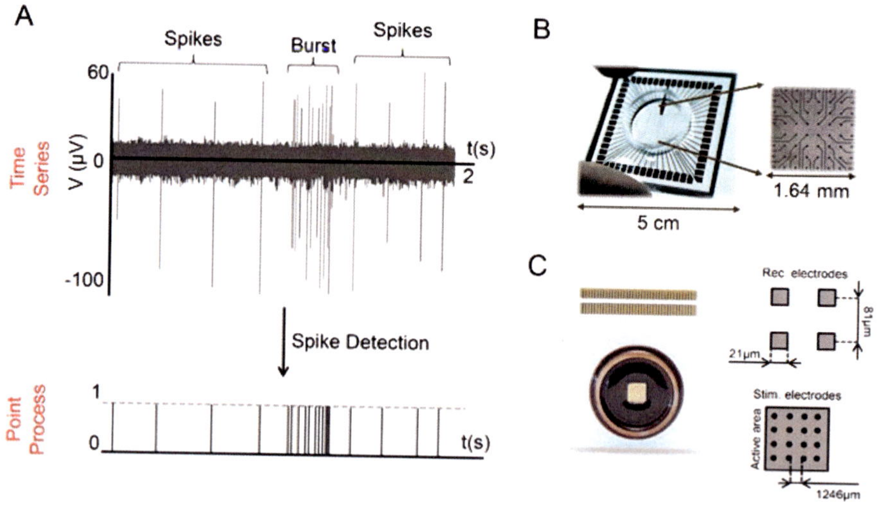

Fig. 1.2 MEA and extracellular signals: **a** The activity of a cortical neural network (28 DIVs) presents a mix of bursting and spiking activity (top). Applying a spike detection algorithm, time series are converted into a serial point process (bottom). **b–c** Examples of Micro-Electrode Arrays (MEAs) made up of **b** 60, **c** 4096 electrodes. Reproduced from [8]

high-density configurations with thousands of microelectrodes (4000–10,000) with a spatial resolution of some tens of micrometers (Fig. 1.2c) [11, 12].

The characteristics of these devices allow different studies on neuronal networks like electrical [13] and chemical manipulation [14], and/or physical segregation in sub-populations (e.g. [15]).

More recently the scientific community is beginning to use MEAs for characterizing the underlying functional connectivity, and its interplay with the expressed dynamics [16], especially by exploiting the high-density systems which allow a more accurate reconstruction of the network topology [17].

1.5 How to Infer Functional Connectivity in In Vitro Neural Networks

In vitro networks represent an excellent benchmark for the validation of functional-effective connectivity methods [8, 18]. Indeed, reconstructing the detailed connectivity of a neuronal network from spike data (functional connectivity) is not trivial, and it is still an open issue, due to the complexities introduced by neuron dynamics and the high number of network connections [19, 20]. Researchers have often studied the status of structural connectivity and its relationship to functional-effective connectivity, and tried to relate connectivity to dynamics [21]. In order to gain understanding of the exchange of information within and between hypothetical neuronal assemblies and to extract topological characteristics, it is necessary to monitor, in parallel, the activities of many neurons and to analyze large amount of data from these electrophysiological measurements. For this purpose, the recent developed high-density MEAs technologies are certainly very attractive since they allow monitoring the ongoing electrophysiological spatio-temporal patterns of complex networks [12, 22], from the majority of neurons in the network.

To estimate the functional-effective connectivity of in vitro networks, there are two different strategies: the first one relies on the direct analysis of the acquired sequence of voltage values (Fig. 1.2a top) from each recording electrode (i.e., the time series). The other approach deals with point processes (e.g., spike trains). Practically, a spike train is a sequence of samples equal to 1 if a spike is detected and 0 otherwise (Fig. 1.1a bottom) [23–26].

Statistical analysis of spike train data was pioneered by Perkel [27] and followed by more than four decades of methodology development in this area [28]. Analytically, cross-correlation based methods remain the main statistics for evaluating interactions among the elements in a neuronal network, and produce a weighted assessment of the connections strength. Weak and non-significant connections may tend to obscure the relevant network topology constituted by strong and significant links, and therefore are often discarded by applying an absolute, or a proportional weighted threshold [29]. Correlation based techniques include: independent components analysis and various measures of synchrony, smoothed ratio of spiking activity [30], cross-correlation and cross-correlogram [31, 32], correlation coefficient [33], partial-correlation [34]. Cross-Correlation (CC) measures the frequency at which one particular neuron or electrode fires ("target") as a function of time, relative to the firing of an event in another network ("reference"). Partial Correlation (PC) is able to distinguish between direct and indirect connections by removing the portion of the relationship between two spike trains that can be attributed to linear relationships with recorded spike trains from other neurons [34, 35]. Other popular techniques to infer functional-effective connectivity are based on Information Theory (IT) methods [36, 37], Granger causality and dynamical causal modeling [38]. Commonly used information theoretic measures able to estimate causal relationships are transfer entropy [39] and joint entropy [18]. Transfer Entropy (TE) is an information-theory based method that estimates the part of the activity of a neuron which depend from the past activity of another neuron. Joint Entropy (JE) is an entropic measure of the cross Inter-Spike Intervals (cISI).

In my PhD work, I focused on CC and Cross-Correlation Histogram (CCH) based methods, PC, TE and JE. With few exceptions [40, 41], all the recently introduced and revisited methods deal with excitation, ignoring inhibition or admitting the failure in reliably identifying inhibitory links [39]. To overcome such a limitation, I introduced a new CCH-based algorithm able to efficiently and accurately infer effective excitatory-inhibitory links (cf. *Filtered Normalized Cross-Correlation Histogram (FNCCH)* section).

In fact, the cross-correlation is able by definition to detect inhibition, but, from some experimental works related to the analysis of cortical connectivity from in vivo multi-unit recordings, it was shown that the sensitivity for excitation is much higher than the sensitivity for inhibition [42] (due to the low firing rates of neurons). A general lower sensitivity of cross-correlation for inhibition vs. excitation has also been proved theoretically [43] thus making the task of inhibitory connection's identification particularly difficult. However, by using my ad-hoc developed FNCCH algorithm, I could derive effective connectivity maps (both for excitation and inhibition) reliably extracting topological characteristics from multiple spike trains in large-scale networks (i.e., thousands of neurons) monitored by large-scale MEAs (i.e., with thousands of micro-transducers).

Relative to the information theory-based methods, I implemented customized versions of TE and JE. In particular, I studied a new version of the TE algorithm, in which I extended the method to deal with multiple time delays (temporal extension) and with multiple binary patterns (high order extension), obtaining improved reliability-precision and increasing computational performances. I also worked on the JE algorithm; in fact, in its original definition the JE provides only a value that estimates the probability of two neurons to be functionally connected, but in my work I added the possibility to extract the temporal delay that characterizes a detected functional connection. It is worth to consider that every connectivity method provides a full $n \ x \ n$ connectivity map, if n is the number of the analyzed electrodes. Thus, a thresholding procedure is required to throw away those values that are close to or in the noise, and not real connections. This requires setting a threshold for the connectivity matrix. Exploring the available works in the literature about the analysis of functional connectivity of in vitro neural networks, it is possible to see several thresholding procedures, with different levels of complexity. The simplest of such procedures is to use a hard threshold [34] defined as ($\mu + n \cdot \sigma$), where μ and σ are the mean and the standard deviation computed among all the CM's elements, respectively, and n is an integer. Another possibility is to use shuffling techniques, that allow to destroy the information stored in the spike timing, obtaining independent spike trains (i.e., surrogate data). In my Ph.D., I focused on both the typologies of thresholding procedures, introducing a customized shuffling approach for the JE algorithm. Such an approach, required high computational efficiency and was made possible only by the optimization strategies that I adopted for the implementation of the connectivity methods.

1.6 Functional Connectivity Analysis Requires Ad-Hoc Software and Algorithms Implementation

The analysis of the interactions in large-scale neuronal networks to infer functional connectivity is a demanding computational process. Recent advances in multichannel extracellular recording techniques have made possible the simultaneous recording of the electrophysiological activity of thousands of neurons, with in vitro arrays that can contain up to thousands of microelectrodes [7, 11, 22, 44, 45]. Such relevant advancements in the technology have made routine the acquisition of massive amounts of data. Moreover, quite often the experimental protocols researchers perform require long recordings (e.g., tens of minutes, hours) which generate a huge amount of raw data. Furthermore, to validate an experimental finding, several trials of the same protocol are necessary, or in other cases, several experimental conditions have to be tested. For these reasons, the three main features that software tools have to satisfy are: (i) reasonable times of computation (efficiency); (ii) simple procedures to process several experiments (automation); (iii) possibility to analyze data coming from different acquisition systems (flexibility). Indeed, the two first conditions should not compromise on the accuracy and the efficacy of the developed algorithms. For the above reasons, new computational strategies are requested for optimizing the management and analysis of the acquired data.

I developed a new software package, named SPICODYN, which implements, starting from raw data, several algorithms to characterize the spiking and bursting properties of large-scale neuronal networks. Such a tool is independent from the acquisition system and is conceived for the analysis of multi-site neuronal spike signals in HDF5 format (HDFGroup 2013). Particularly, it is specifically tailored to process recordings acquired by the commercial High-Density Multi-Electrode Arrays (HD-MEAs) (e.g., 4096 microelectrodes from 3 Brain or 4225 microelectrodes from Multi Channel Systems, MCS, Reutlingen, Germany).

1.7 Goal of the Thesis

Goal of my Ph.D. work has been to develop a group of statistical measures to infer functional connectivity in in vitro neural networks. I worked on correlation-based methods (cross-correlation and partial correlation), and information-theory based methods [Transfer Entropy (TE) and Joint Entropy (JE)]. More in detail, by means of the developed methods, I tried to study the interplay between dynamics and connectivity, using high density resolution systems, with thousands of microelectrodes. To fulfill such an aim, I re-adapted the computational logic operations of the aforementioned connectivity methods to reduce the time requested for the statistical computations. Moreover, I worked on a new method based on the cross-correlogram, able to detect both inhibitory and excitatory links. The FNCCH shows a very high precision in detecting both inhibitory and excitatory functional links when applied to a computational model of neuronal networks. Concerning the information-theory based algorithms, I worked on the temporal and pattern extension of the already existing TE algorithm, by developing the Delayed TE (DTE) and the Delayed High Order TE (DHOTE) algorithms. Finally, starting from the mathematical definition in [18] of the JE, I developed a new customized version of the algorithm capable to detect the delay associated to a functional link, together with a customized shuffling based thresholding approach. Finally, I embedded all of these connectivity methods into a user-friendly open source software named SPICODYN[46] which inherits the backbone structure from TOOLCONNECT[47]. SPICODYN allows the user to perform a complete analysis on data acquired from any acquisition system, including: raw data viewing, spike and burst detection and analysis, functional connectivity analysis, graph theory and topological analysis.

1.8 Structure of the Thesis

In the first section, Materials and methods, I will describe at first the statistical measures that I used to infer functional connectivity in in vitro neural networks coupled to the MEAs. Then, I will introduce the problem of thresholding the connectivity matrix, together with the description of possible solutions that I studied during my work. Finally, I will describe the metrics that I used to evaluate the connectivity methods' performances and all the graph theory's mathematical instruments fundamental to treat the connectivity matrix as a graph, inferring and characterizing the neural networks' topological features. The section Results is divided into two subsections. In the first one, I will describe the results obtained by applying each of the implemented connectivity methods on computational models to evaluate the computational accuracy, as well as an application to experimental neural networks coupled to both low and high density MEAs. In the second one, I will describe the logics of optimization that I adopted in order to reduce the requested time for computation and the software package SpiCoDyn, explaining the informatics tools and implementation choices that allowed an improvement of the computational efficiency, making possible to perform a complete analysis of neural networks coupled to high density acquisition systems. Finally, a last section summarizes all the obtained results and suggests a possible perspective for further extension of my work.

References

1. Makarov VA, Panetsos F, de Feo O (2005) A method for determining neural connectivity and inferring the underlying network dynamics using extracellular spike recordings. J Neurosci Methods 144:265–279. https://doi.org/10.1016/j.jneumeth.2004.11.013
2. Yuste R (2015) From the neuron doctrine to neural networks. Nat Rev Neurosci 16:487–497
3. Sporns O, Tononi G, Edelman GM (2000) Connectivity and complexity: the relationship between neuroanatomy and brain dynamics. Neural Netw 13:909–922
4. Sporns O (2011) The human connectome: a complex network. Ann N Y Acad Sci 1224:109–125. https://doi.org/10.1111/j.1749-6632.2010.05888.x
5. Sporns O, Tononi G, Kotter R (2005) The human connectome: a structural description of the human brain. PLoS Comput Biol 1:e42. https://doi.org/10.1371/journal.pcbi.0010042
6. Sporns O, Tononi G (2002) Classes of network connetivity and dynamics. Complexity 7:28–38
7. Buchs P-A, Muller D (1996) Induction of long-term potentiation is associated with major ultrastructural changes of activated synapses. Proc Natl Acad Sci USA 93:8040–8045
8. Poli D, Pastore VP, Massobrio P (2015) Functional connectivity in *in vitro* neuronal assemblies. Front Neural Cir 9:57. https://doi.org/10.3389/fncir.2015.00057
9. Letourneau PC (1975) Possible roles of cell to substratum adhesion in neuronal morphogenesis. Dev Biol 44:77–91

10. Potter SM, DeMarse TB (2001) A new approach to neural cell culture for long-term studies. J Neurosci Methods 110:17–24
11. Berdondini L et al (2009) Active pixel sensor array for high spatio-temporal resolution electrophysiological recordings from single cell to large scale neuronal networks. Lab Chip 9:2644–2651
12. Frey U, Egert U, Heer F, Hafizovic S, Hierlemann A (2009) Microelectronic system for high-resolution mapping of extracellular electric fields applied to brain slices. Biosens Bioelectron 24:2191–2198. https://doi.org/10.1016/j.bios.2008.11.028
13. Wagenaar DA, Madhavan R, Pine J, Potter SM (2005) Controlling bursting in cortical cultures with closed-loop multi-electrode stimulation. J Neurosci 25:680–688. https://doi.org/10.1523/JNEUROSCI.4209-04.2005
14. Pancrazio JJ et al (2003) A portable microelectrode array recording system incorporating cultured neuronal networks for neurotoxin detection. Biosens Bioelectron 18:1339–1347
15. Levy O, Ziv NE, Marom S (2012) Enhancement of neural representation capacity by modular architecture in networks of cortical neurons. Eur J Neurosci 35:1753–1760. https://doi.org/10.1111/j.1460-9568.2012.08094.x
16. Massobrio P, de Arcangelis L, Pasquale V, Jensen HJ, Plenz D (2015) Criticality as a signature of healthy neural systems. Front Syst Neurosci 9:22. https://doi.org/10.3389/fnsys.2015.00022
17. Maccione A et al (2012) Multiscale functional connectivity estimation on low-density neuronal cultures recorded by high-density CMOS micro electrode arrays. J Neurosci Methods 207:161–171. https://doi.org/10.1016/j.jneumeth.2012.04.002
18. Garofalo M, Nieus T, Massobrio P, Martinoia S (2009) Evaluation of the performance of information theory-based methods and cross-correlation to estimate the functional connectivity in cortical networks. PLoS ONE 4:e6482. https://doi.org/10.1371/journal.pone.0006482
19. Van Bussel F, Kriener B, Timme M (2011) Inferring synaptic connectivity from spatio-temporal spike patterns. Front Comput Neurosci 5:3. https://doi.org/10.3389/fncom.2011.00003
20. Doiron B, Litwin-Kumar A, Rosenbaum R, Ocker GK, Josic K (2016) The mechanics of state-dependent neural correlations. Nat Neurosci 19:383–393. https://doi.org/10.1038/nn.4242
21. Sporns O (2013) Structure and function of complex brain networks. Dialogues Clin Neurosci 15:247–262
22. Eversmann B et al (2003) A 128 x 128 CMOS biosensor array for extracellular recording of neural activity. IEEE J Solid-State Circ 38:2306–2317
23. Maccione A et al (2009) A novel algorithm for precise identification of spikes in extracellularly recorded neuronal signals. J Neurosci Methods 177:241–249
24. Ide AN, Andruska A, Boehler M, Wheeler BC, Brewer GJ (2010) Chronic network stimulation enhances evoked action potentials. J Neural Eng 7:016008. https://doi.org/10.1088/1741-2560/7/1/016008
25. Egert U et al (2002) MEA-Tools: an open source toolbox for the analysis of multi-electrode data with MATLAB. J Neurosci Methods 117:33–42
26. Borghi T, Gusmeroli R, Spinelli AS, Baranauskas G (2007) A simple method for efficient spike detection in multiunit recordings. J Neurosci Methods 163:176–180
27. Gerstein GL, Perkel DH (1969) Simultaneously recorded trains of action potentials: analysis and functional interpretation. Science 164:828–830
28. Brown EN, Kass RE, Mitra PP (2004) Multiple neural spike train data analysis: state-of-the-art and future challenges. Nat Neurosci 7:456–461
29. Friston KJ (2009) Modalities, modes and models in functional neuroimaging. Science 326:399–403. https://doi.org/10.1126/science.1174521
30. Ventura V, Cai C, Kass RE (2005) Statistical assessment of time-varying dependency between two neurons. J Neurophysiol 94:2940
31. Brosch M, Schreiner CE (1999) Correlations between neural discharges are related to receptive field properties in cat primary auditory cortex. Eur J Neurosci 11:3517–3530. https://doi.org/10.1046/j.1460-9568.1999.00770.x
32. Salinas E, Sejnowski TJ (2001) Correlated neuronal activity and the flow of neural information. Nat Rev Neurosci 2:539–550

33. Bedenbaugh P, Gerstein GL (1997) Multiunit normalized cross correlation differs from the average single-unit normalized correlation. Neural Comput 9:1265–1275
34. Poli D, Pastore VP, Martinoia S, Massobrio P (2016) From functional to structural connectivity using partial correlation in neuronal assemblies. J Neural Eng 13:026023
35. Eichler M, Dahlhaus R, Sandkuhler J (2003) Partial correlation analysis for the identification of synaptic connections. Biol Cybern 89:289–302
36. Quiroga QR, Panzeri S (2009) Extracting information from neuronal populations: information theory and decoding approaches. Nat Rev Neurosci 10:173–185
37. Nigam S et al (2016) Rich-club organization in effective connectivity among cortical neurons. J Neurosci 36:670–684. https://doi.org/10.1523/JNEUROSCI.2177-15.2016
38. Friston KJ (2011) Functional and effective connectivity: a review. Brain Connectivity 1:13–36. https://doi.org/10.1089/brain.2011.0008
39. Ito S et al (2011) Extending Transfer entropy improves identification of effective connectivity in a spiking cortical network model. PLoS ONE 6:e27431. https://doi.org/10.1371/journal.pone.0027431
40. Freeman GM Jr, Krock RM, Aton SJ, Thaben P, Herzog ED (2013) GABA networks destabilize genetic oscillations in the circadian pacemaker. Neuron 78:799–806. https://doi.org/10.1016/j.neuron.2013.04.003
41. Zaytsev YV, Morrison A, Deger M (2015) Reconstruction of recurrent synaptic connectivity of thousands of neurons from simulated spiking activity. J Comput Neurosci 39:77–103. https://doi.org/10.1007/s10827-015-0565-5
42. Abeles M (1991) Corticonics: neural circuits of the cerebral cortex. Cambridge University Press, Cambridge
43. Aertsen A et al (1991) Neural interactions in the frontal cortex of a behaving monkey: signs of dependence on stimulus context and behavioral state. J Hirnforsch 32:735–743
44. Muller J et al (2015) High-resolution CMOS MEA platform to study neurons at subcellular, cellular, and network levels. Lab Chip 15:2767–2780. https://doi.org/10.1039/c5lc00133a
45. Viswam V, Dragas J, Muller J, Hierlemann A In: IEEE international solid-state circuits conference. IEEE, pp 394–396
46. Pastore VP, Godjoski A, Martinoia S, Massobrio P (2017) SPICODYN: a toolbox for the analysis of neuronal network dynamics and connectivity from multi-site spike signal recordings. Neuroinformatics 1:15–30. https://doi.org/10.1007/s12021-017-9343-z
47. Pastore VP, Poli D, Godjoski A, Martinoia S, Massobrio P (2016) ToolConnect: a functional connectivity toolbox for *in vitro* networks. Front Neuroinf https://doi.org/10.3389/fninf.2016.00013

Chapter 2
Materials and Methods

2.1 Correlation-Based Connectivity Methods

2.1.1 Cross-Correlation

Cross-correlation [1] measures the frequency at which one particular neuron or electrode fires ("target") as a function of time, relative to the firing of an event in another one ("reference"). Mathematically, the correlation function is a statistic representing the average value of the product of two random processes (the spike trains). Given a reference electrode x and a target electrode y, the correlation function reduces to a simple probability $C_{xy}(\tau)$ of observing a spike in one train y at time $(t + \tau)$, given that there was a spike in a second train x at time t; τ is called the time shift or the time lag. For my Ph.D. studies, I used the standard definition for the cross-correlation computation, following a known normalization approach on the *CC* values [2]. We can define the cross-correlation as follows:

$$C_{xy}(\tau) = \frac{1}{\sqrt{N_x N_y}} \sum_{s=1}^{Nx} x(t_s) y(t_s - \tau) \tag{2.1}$$

where t_s indicates the timing of a spike in the x train, N_x is the total number of spikes in the x train and N_y is the total number of spikes in the y train. Cross-correlation is limited to the interval [0, 1] and is symmetric $C_{xy}(\tau) = C_{yx}(-\tau)$. The cross-correlogram is then defined as the correlation function computed over a chosen correlation window (W, $\tau = [-W/2, W/2]$). Figure 2.1 describes a possible strategy to compute the Cross-Correlation Histogram (CCH). The Normalized CCH (NCCH) is obtained by applying the normalization factor represented by the squared product of the target and the reference number of spikes to the CCH. Different shapes of cross-correlograms can be obtained from pairs of analyzed spike trains. When the two spike trains are independent, the cross-correlogram is flat; if there

V. P. Pastore, *Estimating Functional Connectivity and Topology in Large-Scale Neuronal Assemblies*, Springer Theses, https://doi.org/10.1007/978-3-030-59042-0_2

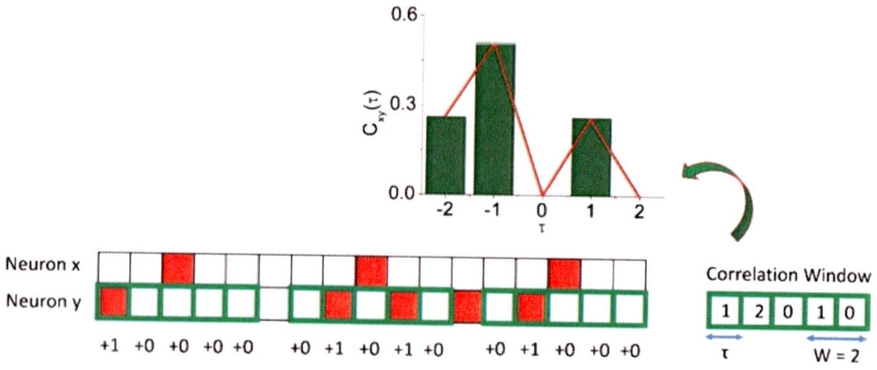

Fig. 2.1 Computation of the cross-correlation histogram. The cross-correlation histogram (CCH) is computed according to a discrete convolution. Thus, the correlation window is centered in each of the reference train spikes (red squares). The number of target spikes inside each correlation window's bin (green squares) are counted, and the cumulative number of spikes, after normalization, represent the cross-correlation histogram

is any covariation in the spike trains, one or more peaks appear (Fig. 2.2c). The occurrence of significant departures from a flat background in the cross-correlogram (i.e., a peak or a trough) is an indication of a functional connection. In particular, a peak corresponds to an excitatory connection and a trough is relative to an inhibition. The different amplitude of the peaks can be related to the existence of different levels of correlation between neural spike trains. A peak can appear also for other kinds of interaction (e.g., covariations over in response latency, and in neuronal excitability). Generally, a correlogram can reflect a so-called direct excitatory connection between the two neurons when a one-sided peak is evident (Fig. 2.2c). This peak is displaced from the origin of time by latency corresponding to the synaptic delay.

2.1.2 *Cross-Correlation Histogram*

The use of spike trains data offers the possibility to optimize the cross-correlation algorithm efficiency. Thus, to overcome the lack of efficiency of many of the proposed cross-correlation computation strategies, here I present an alternative approach based on the "direct" spike time stamps inspection that avoids un-necessary calculations on the binarized spike trains. The only important information is stored in the bins containing a spike (i.e., spike time stamp), that are significantly less than null bins. If we consider that, typically, the average mean firing rate in neural networks oscillates between 0.2 spikes/s and 20 spikes/s [3], acquiring with a sampling frequency of 10 kHz means we will have only 2% of bin with spikes. To account only for the spike trains time stamps, I developed a new logical version of the CCH.

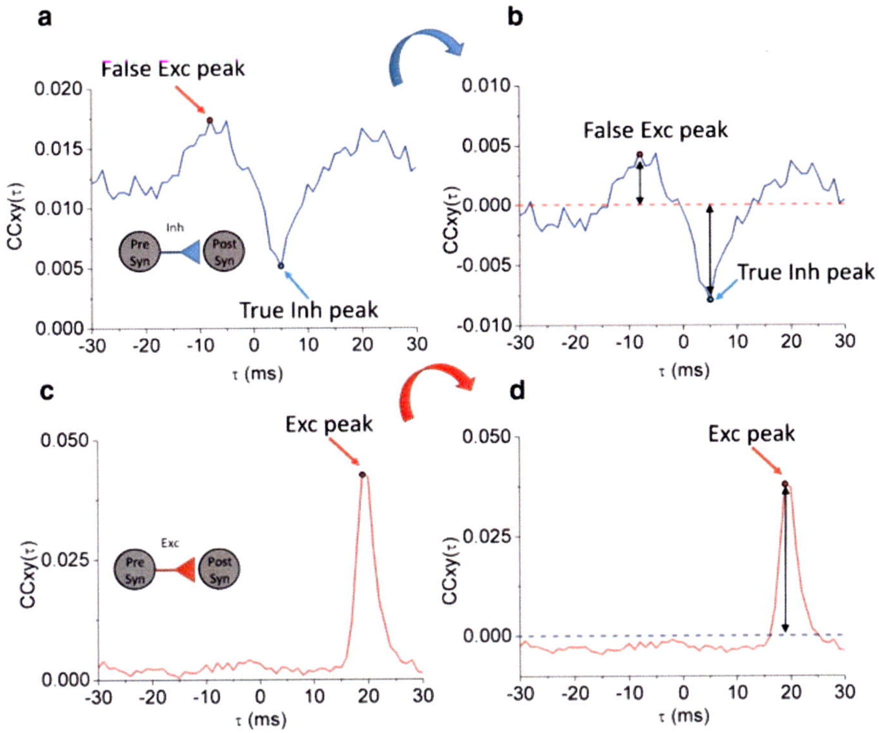

Fig. 2.2 Example of FNCCH detection for excitatory and inhibitory links in a network model. **a** NCCH computed between two spike trains correspondent to two neurons linked by an inhibitory link in the model. The NCCH might detect a false excitatory peak (blue circle). **b** FNCCH of the two neurons of panel c. The filtering procedure allows to recognize the trough and to detect the negative peak correspondent to the inhibitory link (blue circle). **a** NCCH computed between two spike trains correspondent to two neurons linked by an excitatory link in the model (identified by a red circle). **b** FNCCH of the two neurons of panel a. The "entity peak" allows a better recognition of the excitatory link. Adapted from [4]

2.1.3 Filtered and Normalized Cross-Correlation Histogram (FNCCH)

Let us consider a reference neuron x and a target neuron y, and let us suppose that we computed the NCCH between x and y. After the NCCH computation, the maximum value (i.e., the peak) is commonly used as a value reflecting the strength of the estimated functional link. If x and y share an excitatory link this procedure works well (Fig. 2.2a). When x inhibits y, instead, the inhibitory trough will be discarded in favor of the NCCH peak (Fig. 2.2c), with a misleading excitatory link detection.

Equation (2.2) gives the mathematical definition of the FNCCH computation, that overcomes this problem.

$$FNCCH_{xy\_peak} = C_{xy}(t)\big|_{t} = \arg\max_{t}\left|C_{xy}(t) - \frac{1}{W}\sum_{v=-\frac{W}{2}}^{v=\frac{W}{2}} C_{xy}(v)\right| \tag{2.2}$$

I refer to the filtered peak value as entity of the peak. In this way, it is possible to distinguish between peaks and troughs by taking into account the sign. A negative peak is referred to an inhibitory link (Fig. 2.2b), while conversely a positive peak is referred to an excitatory link (Fig. 2.2d). I also implemented and applied a post computation filtering procedure to improve the detectability of inhibitory links on noisy spike trains.

2.1.3.1 Post Computation FNCCH Filtering

The temporal occurrence of the extracted peak of the NCCH represents the time delay of the identified connection between two neurons. When we deal with experimental recordings, the NCCH could be very jagged, with a shape characterized by oscillations in the central region of the correlation window and a typical decrease of the NCCH values in the borders. The FNCCH computation procedure permits to extract the peaks evaluating their sign, but it is sensitive to a decrease in synchrony due to uncorrelated activity. Such a decrease appears at the boundaries of the correlation window (Fig. 2.3, black curves), and can be exchanged for a decrease in synchrony related to an inhibitory connection. Thus, we can expect that in specific cases characterized by a very jagged correlogram (e.g., due to a low firing rate), this procedure

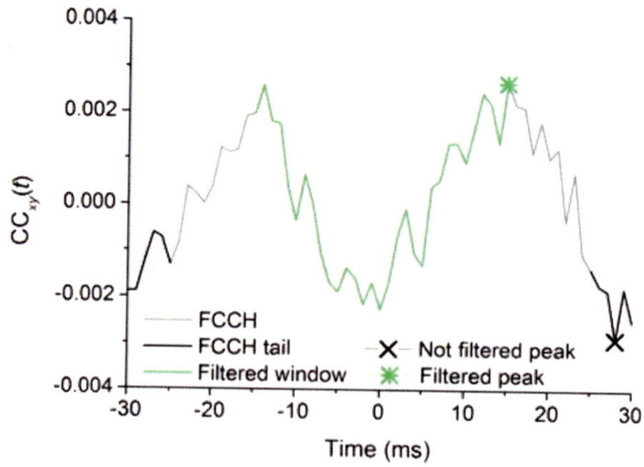

Fig. 2.3 FNCCH post filtering procedure. In this illustrative case, correspondent to weak correlation, the filtering procedure infers a negative value in the boundary region of the correlation window (black line) leading to a false positive inhibitory link. To avoid this, heuristic post filtering procedure is formed by a peak search re-applied in a smaller region of the correlation window (green line) discarding part of the tail. The resulting peak, in this example, is excitatory and with a shorter delay. Reproduced from [4]

will introduce some mistakes in the inhibitory connections detection. I defined this artifact as *tail effect* on the FNCCH. For this reason, I implemented a post FNCCH filtering operation that allows to account for the presence of these artifacts removing them from the set of identified inhibitory connections. More specifically, the filtering procedure (Fig. 2.3) consists in few steps applied to every negative FNCCH values falling in one of the two boundary regions of the correlation window (defined as the 15% of half the amplitude of the correlation window, Fig. 2.3 black curves). If such detected inhibitory link is an artifact due to the tail effect, the FNCCH values in the boundary region will be all negative and likely with a decreasing trend. In this case, the corresponding putative peak is discarded and a new search starts for another peak excluding the boundary regions. We then search for a peak in the central region of the correlation window that I called "peak re-computation window" [Fig. 2.3, green curve, by applying Eq. (2.1)]. Indeed, the "new" peak can be related either to an inhibitory effective connection or to an excitatory one. On the other hand, if we find an increase in the FNCCH reaching a positive value before the end of the correlation window, we can state that the negative peak is correctly identified (i.e., an actual minimum).

2.1.4 Partial Correlation

Cross-correlation based methods are not able to distinguish between direct and indirect connections. In order to overcome this limitation it was introduced the notion of partial coherence [5] in which the effects of the activity of all other spike trains (assumed to be additive) could be removed. Eichler [6], instead, presented a partialization method in the time domain, based on a scaled version of the partial covariance density known as Scaled Partial Covariance Density (SPCD). The SPCD combines the advantages of the cross-correlation histograms and the partialization analysis in the frequency domain: (i) it interprets, in the same way of the cross-correlation histograms, peaks and troughs as excitatory and inhibitory connections respectively; (ii) it allows to discriminate direct and indirect connections and common inputs (see Fig. 2.4).

In general, let us consider a neuronal population V and two specific neurons x, $y \in V$. Let $R_{xy}(\tau)$ be their Standard Correlation, and $R_{xx}(\tau)$ and $R_{yy}(\tau)$ the Auto-Correlation of x and y respectively [7]. The Fourier Transform of $R_{xy}(\tau)$, i.e. the cross-spectral density $S_{xy}(\omega)$, will define the Spectral Coherence $C_{xy}(\omega)$ as follows:

$$C_{xy}(\omega) = \frac{S_{xy}(\omega)}{\sqrt{S_{xx}(\omega)S_{yy}(\omega)}} \tag{2.3}$$

where $S_{xx}(\omega)$ and $S_{yy}(\omega)$ are the Fourier Transform of $R_{xx}(\tau)$ and $R_{yy}(\tau)$, respectively.

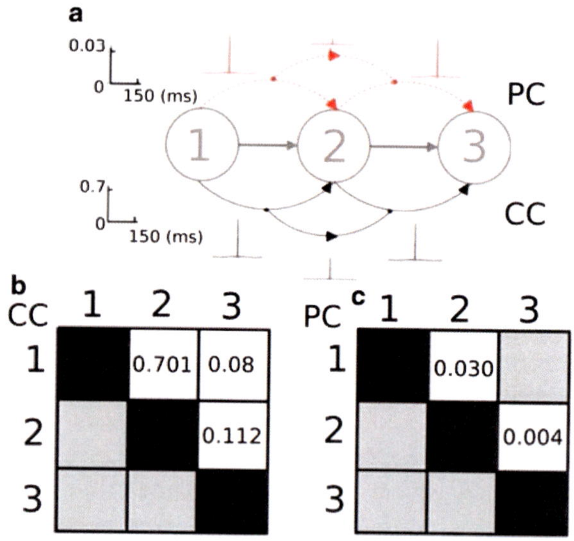

Fig. 2.4 Schematic representation of the partialization process. CM built using cross-correlation (**b**) and partial correlation (**c**) for a simple network made up of 3 neurons, **a** correspondent sketch of connections. PC (differently from CC) does not detect the indirect link between neuron 1 and neuron 3 (yellow circle)

The partialization process [5] removes from $S_{xy}(\omega)$ the effect Z related to all other (possibly multivariate) spike trains of the population V in the following way:

$$Z = V - [x, y] \tag{2.4}$$

$$S_{xy|Z}(\omega) = S_{xy}(\omega) - S_{xZ}(\omega)S_{ZZ}^{-1}(\omega)S_{Zy}(\omega) \tag{2.5}$$

where $S_{ZZ}(\omega)$ is the auto-correlation of Z in the frequency domain and the Inverse Fourier transform of $S_{xy|Z}(\omega)$, $R_{xy|Z}(t)$, is the partial covariance density. Basically, the partialization process consists in iterating the cross-spectral power density computation by removing a node per time, subtracting the new computed value from the cross-spectral power density correspondent to the couple (x, y). If there is the influence of a third node, that is, if the investigated connection is likely to be indirect, such an operation will significantly modify the cross-spectrum value.

$C_{xy|Z}(\omega)$ corresponds to the partial coherence function. It can be defined by the inversion of the spectral matrix $S(\omega)$ of the whole set of nodes [6, 8]. If $G(\omega) = S(\omega)^{-1}$, then it can be mathematically proved, that the partial auto spectrum densities correspond to:

$$S_{xx|V\setminus\{x\}}(\omega) = \frac{1}{G_{xx}(\omega)} \tag{2.6}$$

$$S_{yy|V\setminus\{y\}}(\omega) = \frac{1}{G_{yy}(\omega)} \tag{2.7}$$

Moreover, it can be proved that the partial coherence function can be expressed in function of $G(\omega)$ as:

$$C_{xy|Z}(\omega) = -\frac{G_{xy}(\omega)}{\sqrt{G_{xx}(\omega)G_{yy}(\omega)}} \tag{2.8}$$

We can use the partial coherence function to compute the partial power spectral density $S_{xy|Z}(\omega)$:

$$S_{xy|Z}(\omega) = \frac{C_{xy|Z}(\omega)}{1 - \left|C_{xy|Z}(\omega)\right|^2} \tag{2.9}$$

And substituting Eqs. (2.6), (2.7) and (2.8) into Eq. (2.9), it is possible to obtain:

$$S_{xy|Z}(\omega) = -\frac{G_{xy}(\omega)}{\sqrt{G_{xx}(\omega)G_{yy}(\omega)}}\frac{C_{xy|Z}(\omega)}{1 - \left|C_{xy|Z}(\omega)\right|^2}\sqrt{S_{xx|V\setminus\{x\}}(\omega)S_{yy|V\setminus\{y\}}(\omega)} \tag{2.10}$$

The partial covariance density $R_{xy|Z}(t)$ corresponds to the Fourier inverse transform of the partial power spectral density. The partial correlation function (or SPCD) that I used in my Ph.D.'s work is a scaled version defined of $R_{xy|Z}(t)$ defined as:

$$s_{xy|Z}(t) = \frac{R_{xy|Z}(t)}{\sqrt{r_x r_y}} \tag{2.11}$$

where r_x and r_y are the maximum peak values of the autocorrelation function. Finally, PC function, as well as CC, permits to recognize the directionality of the connections by observing the peak latency from zero.

2.2 Information-Theory Based Connectivity Methods

2.2.1 *Transfer Entropy*

Transfer Entropy (TE) is an asymmetric directed information theoretic measure which allows to extract causal relationships from time series [9], estimating the part of the activity of one single neuron which does not depend only on own past, but also by neural past activity of another cell. In other words, TE measures the flow of information between the activity of two cells. Let us consider a reference spike train y, and a target spike train x. The mathematical definition of transfer entropy is given by Eq. (2.12):

$$TE_{y \geq x} = \sum_t p(x_{t+1}, x_t, y_{t+1}) \log \frac{p(x_{t+1}|x_t, y_{t+1})}{p(x_{t+1}, x_t)} \tag{2.12}$$

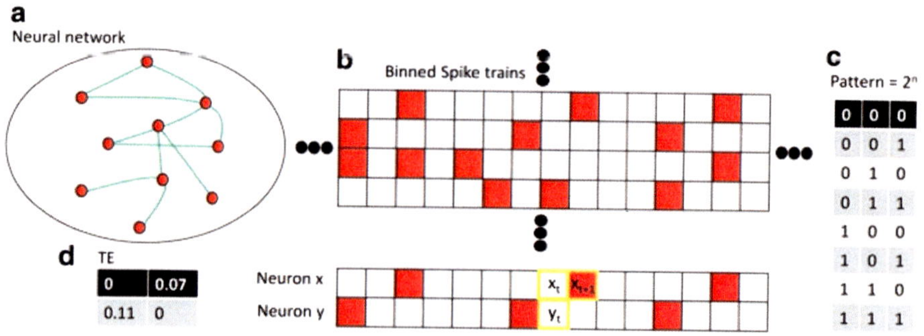

Fig. 2.5 Schematic representation of the Transfer entropy computation process. **a** Recorded in vitro neural networks or in silico model from which binned spike trains (**b**) are extracted. Considering the reference neuron y and the target neuron x (**b**, bottom), Eq. (2.12) is applied by extracting all the possible patterns (**c**) for the binary trains, obtaining the TE's values (**d**)

where x_{t+1} is the bin representing the present activity of the reference train, while x_t and y_t are the past activity of the two trains. See Fig. 2.5 for a schematic description of the computation process.

2.2.2 *Delayed Transfer Entropy*

When two neurons are activity-dependent and connected in a causal dependent way, the consequent information flow usually appears in the correspondent spike trains with a specific delay, different for each couple of neurons. If TE's bin is not large enough, it is highly probable to miss the detection of such effective connection. On the other side, if the bin is too large, more spikes collapse into a single bin causing a likely decreasing in the estimated amount of activity dependence. A possibility to overcome these limitations is to temporally extend TE [10, 11]. The Delayed Transfer Entropy (DTE) is defined according to Eq. (2.13):

$$DTE_{y\geq x}(d) = \sum\nolimits_t p(x_{t+1}, x_t, y_{t+1-d}) \log \frac{p(x_{t+1}||x_t, y_{t+1-d})}{p(x_{t+1}, x_t)} \tag{2.13}$$

In particular, we can start to consider the past bins for the target train at a specific temporal distance [parameter d in Eq. (2.12)] before the present bin in the reference train. Considering d varying from 1 to a fixed number, we can build a temporal function TE(d). The estimated connection among the analyzed spike trains can be extracted using the peak of the TE(d) function and/or the Coincidence Index (CI) [12], i.e. the ratio of the integral of the function in a specified area around the maximum peak to the integral of the total area.

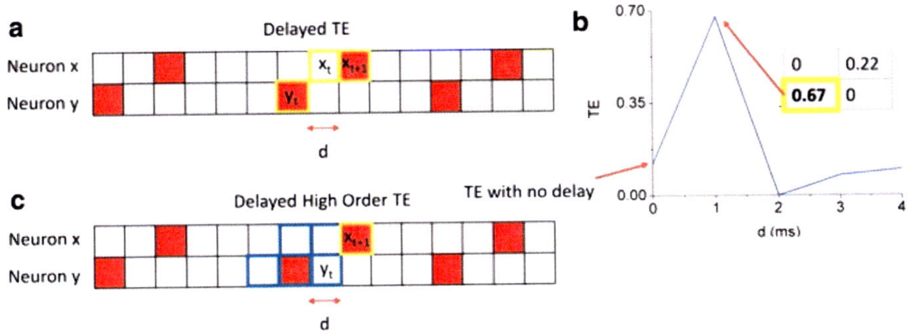

Fig. 2.6 Schematic representation of the DTE and DHOTE computation process. **a** The past bins for the reference train are shifted of a quantity d, that is the parameter of the function DTE(d) (**b**). If multiple bins are considered (blue squares in panel (**c**), the DHOTE is obtained)

2.2.3 *Delayed High Order Transfer Entropy*

It is possible to extend the TE to deal with multiple delays. In Eq. (2.12), the part of past activity to take into account in the TE's computation (number of bins) for the reference train is indicated by k while l represents the number of bin to consider for the target train. Thus, we can define the Delayed High Order Transfer Entropy (DHOTE) according to Eq. (2.14):

$$DHOTE_{y\geq x}(d) = \sum_t p\left(x_{t+1}, x_t^{(k)}, y_{t+1-d}^{(l)}\right) \log \frac{p\left(x_{t+1} | x_t^{(k)}, y_{t+1-d}^{(l)}\right)}{p\left(x_{t+1}, x_t^{(k)}\right)} \tag{2.14}$$

In this way, the couple (k, l) defines the TE's order. For (k, l) different from the couple $(1, 1)$ we define the High Order Transfer Entropy (HOTE). The more is the increase of the TE's order (taking into account more bins in the two trains) the more precise is the estimation of the information flow with a subsequent increasing of the time required for the computation. See Fig. 2.6 for a schematic description of the computation process.

2.2.4 *Joint Entropy*

The JE algorithm is based on the computation of the cross-Inter Spike Intervals (cISI). It is based on the idea that if two spike trains are correlated, their activity will show a specific degree of synchrony. Let us introduce a reference spike train x and a target spike train y. We can define the cISI as:

$$cISI = t_x - t_y \tag{2.15}$$

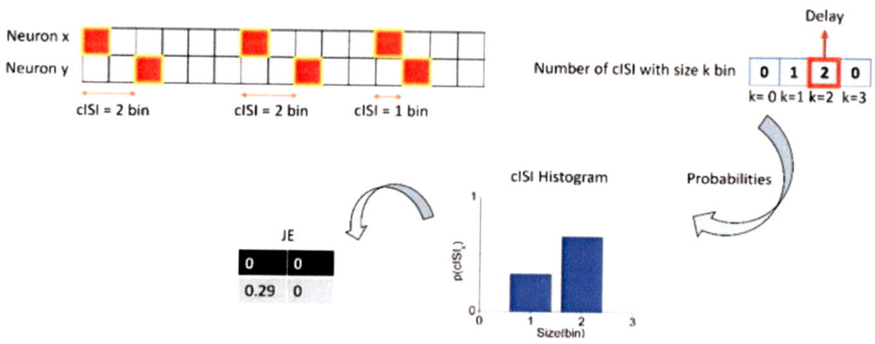

Fig. 2.7 Schematic representation of the JE computation process. The cISIs between reference and target trains are computed. Taking into account the occurrences of each cISI it is possible to build the cISI histogram. The application of Eq. (2.16) allows to compute the JE between the two investigated electrodes

where t_x is the time stamp correspondent to a reference spike train, while t_y is the first time stamp temporally following the reference one in the target train. By iterating the cISI computation for every reference spike, we can build the cISI histogram. Thus, if $cISI_k$ corresponds to a cISI of size k bins, we can introduce the JE measure as:

$$JE(x, y) = -\sum p(cISI_k) * \log(p(cISI_k)) \tag{2.16}$$

The higher the probability of two neurons being correlated in their activity, the lower the JE value. In the ideal case of autocorrelation, JE will be equal to 0. The more jagged is the cISI histogram, the less is the correlation of two spike trains, causing the growing of the JE value, according to Eq. (2.16). In its original definition [13], JE provides only a number for each couple of neurons, with no temporal information of the delay of a functional connection.

To overcome such a limitation, I introduced and tested a method for extracting the delay. Such procedure is simply based on extracting the most frequent cISI's size in bins (see Fig. 2.7). This value corresponds to the most likely delay of the investigated functional connection.

2.3 Thresholding Approaches

The statistical measures described in the previous paragraphs enable building a Connectivity Matrix (CM). The CM is a $n \times n$ matrix (where n is the number of analyzed nodes (i.e., electrodes or neurons)) whose generic element (i, j) is the estimation of the strength of connection between electrodes i and j. Any connectivity method provides a value for each couple of analyzed electrodes, even if such electrodes do not share a real functional connection. As a consequence, the CM is a

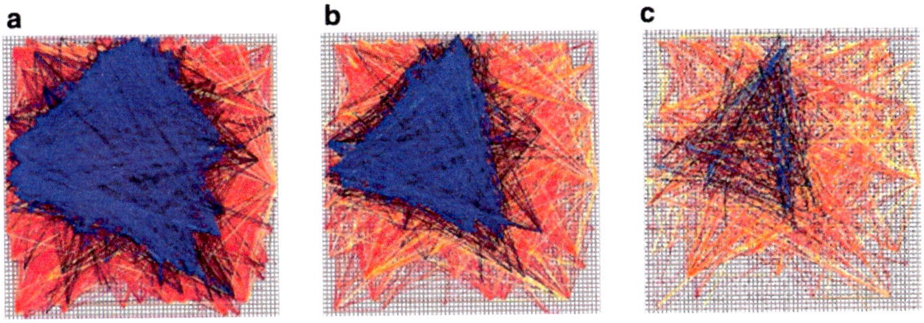

Fig. 2.8 Hard Thresholding dependence on the parameter n. Connectivity graph of a cortex neural network coupled to the MEA-4 k acquisition system, correspondent to: 1 (**a**), 2 (**b**), 3 (**c**)

full matrix of n [14] elements, and a thresholding procedure is required to throw away those values that are not significant since close to the noise level, and then not representing real connections.

2.3.1 Hard Threshold

The simplest thresholding procedure is to apply a hard threshold [15] to the CM. One of the simplest choices is to use a threshold equal to $\mu + n * \sigma$, where n is an integer, while μ and σ are the mean value and the standard deviation of the connectivity matrix elements, respectively [16]. Such a procedure is based on the assumption that the strongest a link, the most likely it corresponds to a true functional link. The main limit of the hard threshold approach is the choice of the parameter n, that is completely arbitrary. A little variation on the threshold (i.e., the value of n) can drastically alter the results of a functional connectivity analysis, providing different graphs with different topological parameters (see Fig. 2.8). Even if the basic assumption on the links strength can be correct, the problem of defining how strong is enough, could still affect the obtained results with a bias dependent on the choice of the parameter n in the thresholding procedure.

2.3.2 Shuffling Approach

Another possible solution to threshold the connectivity matrix is the shuffling thresholding approach [17]. This method is based on the generation of spike trains surrogate data. A surrogate data is generated from a spike train by preserving all the statistical features except for the one that we want to test for the significance (i.e., the timing of the spikes). In fact, the correlation between two spike trains is embedded in the temporal positions and synchrony of the spikes. If we destroy such information, by randomly changing the positions of the target spikes, we can build a null hypothesis, that is correspondent to a non-likely connection. I will describe the shuffling approach that I implemented and tailored for the JE algorithm (see Fig. 2.9).

In detail, after generating surrogate data (see Fig. 2.10), it is possible to test the significance of a found functional connection between spike trains x and y, by applying the JE method to the pair (x, shuffled y) for every surrogate target train. Such JE value is correspondent to a non-connection, thus, we can test if the JE value correspondent to the pair (x, y) is lower than the shuffled one, indicating a likely significant functional connection. If we generate hundreds or thousands of shuffled trains, we can repeat such test a large number of times. Finally, we can define a

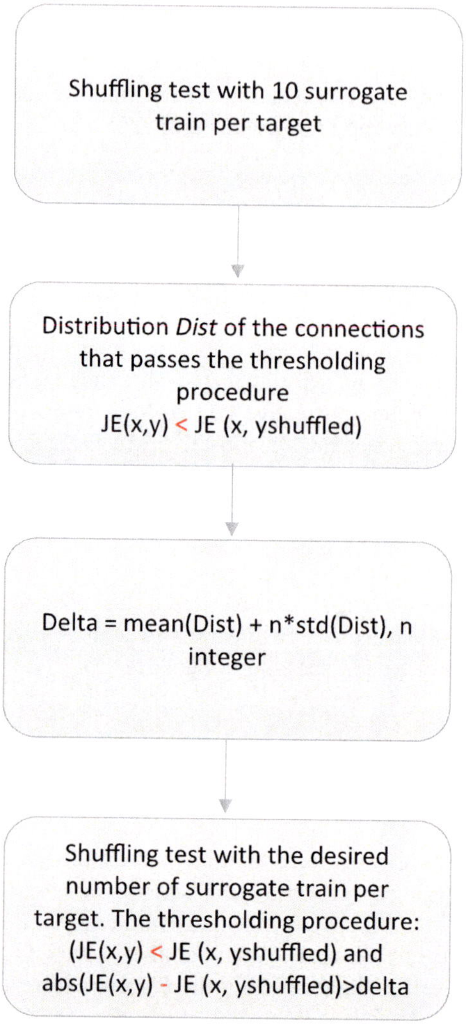

Fig. 2.9 Schematic representation of the customized thresholding approach based on surrogate data generation

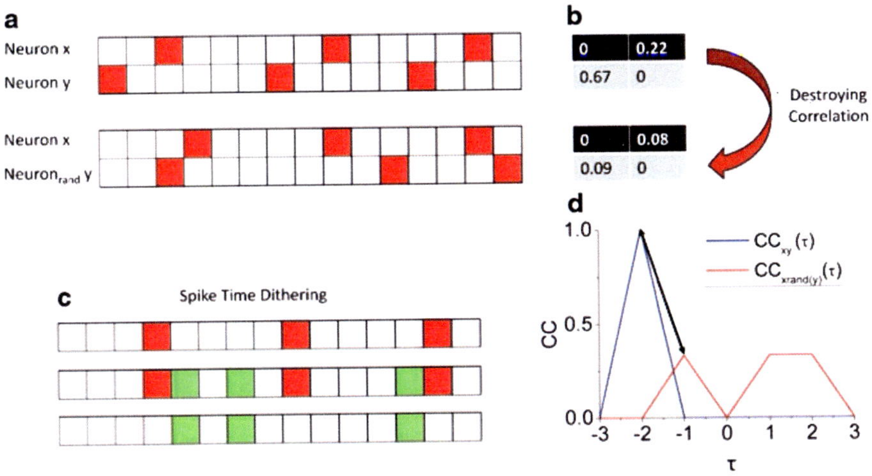

Fig. 2.10 Schematic representation of the shuffling process. If we consider two strongly correlated spike trains x and y as in panel (**a**), they will provide high values for every applied connectivity method. As an example, panel **b** reports the computed DTE's values. If we destroy the correlation, just randomly changing the position of the spike in the neuron y (red squares in all the panels), the detected values decrease (panel (**b**), bottom). The idea behind the shuffling process is to destroy correlation randomly changing the position of the spikes, as reported in panel (**c**) (green squares represent the shuffled spikes). It is possible to shuffle the reference, the target or both the spike trains, comparing the obtained values with the ones corresponding to the original trains. As an example, panel **d** shows how the NCCH correspondent to original trains (blue line) and shuffled target train (red line). We can notice that the shuffled peak is lower than the original one, suggesting that the investigated functional connection is significant. Otherwise, if the shuffled peak is higher or equal to the original one, we can assess that the functional connection is not significant, discarding it

minimum p-value for accepting as significant a functional connection, and apply the thresholding procedure to the entire dataset.

In my work, I introduced a parameter that I called delta, that is a minimum difference between the value representing the null hypothesis and the original statistical one, in order to increase the precision reducing the number of false positives. In detail, I implemented a bootstrap shuffling analysis to extract the parameter delta. Thus, I start a pre-processing analysis performing the shuffling test with 10 shuffled spike trains for each electrode. I compute the distribution of the differences between the original JE value and the shuffled one, among all the statistically significant estimated connections, that is, considering those differences for which the original JE value is lower than the shuffled one. At this stage, I define the delta value as an integer multiple of the mean value of such distribution (see Fig. 2.9).

As we can see in the results, such a parameter hardly influences the performances. In the literature, several methods to generate surrogate data from spike trains can be found. In my work, I followed the spike time dithering approach [17]. The spike time dithering procedure consists in randomly shuffling each spike inside a window of specified width. Thus, each spike at time t is randomly substituted by a spike at a time $\in [t\text{-}w, t + w]$, where w is the shuffling window's width.

2.4 Metrics to Evaluate the Connectivity Methods' Performances

After the thresholding procedure, any connectivity method works as a binary classifier. In fact, the method classifies whether two electrodes$i$ and j share a functional link or not. It is evident that all the metrics used to evaluate a binary classifier can be applied to quantify a connectivity method's computational accuracy. For this purpose, the testing of a connectivity method requires the use of a neural network computational model, and the consequent comparison between prediction and observation. Using such computational models, we have the prediction represented by the computed thresholded connectivity matrix and the observation represented by the model's Synaptic Weight Matrix (SWM). The higher the correspondence between prediction and observation, the higher the computational accuracy of a method.

2.4.1 *Receiver Operating Characteristic (ROC) Curve*

The ROC curve [18] is a common metrics used to evaluate the performances of a binary classifier by comparing prediction and observation. In my PhD study, the prediction is represented by the computed Thresholded Connectivity Matrix (TCM), while the observation corresponds to the SWM of the neural network model (i.e., the ground truth).

We can define the True Positive Rate (TPR) and the False Positive Rate (FPR) as follows:

$$TPR = \frac{TP}{TP + FN} \tag{2.17}$$

$$FPR = \frac{FP}{FP + TN} \tag{2.18}$$

The ROC curve (Fig. 2.11a) is then obtained by plotting TPR versus FPR. The Area Under Curve (AUC) is a main parameter extracted to have a single number describing the performances of a binary classifier: a random guess will correspond to 0.5 (straight line in Fig. 2.11a), while a perfect classifier will have a value of 1. Another important metric that can be extracted from a ROC analysis is the accuracy, defined as:

$$ACC = \frac{TP + TN}{TP + TN + FP + FN} \tag{2.19}$$

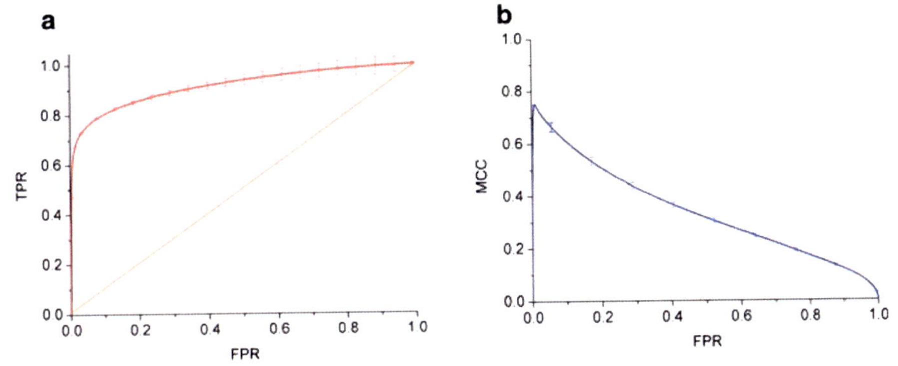

Fig. 2.11 Metrics to evaluate a binary classifier. **a** Example of ROC curve. The random guess corresponds to the ROC curve being a line coincident with the bisector. **b** Example of MCC curve

2.4.2 Matthews Correlation Coefficient (MCC) Curve

The MCC curve [18] is a common metrics, alternative to the ROC analysis, used to evaluate the performances of a binary classifier by comparing prediction and observation. Using the quantities defined in the previous paragraph, changing the threshold used to compute the TCM, we can define the MCC as:

$$MCC = \frac{TP * TN - FP * FN}{\sqrt{(TP + FP)(TP + FN)(TN + FP)(TN + FN)}} \tag{2.20}$$

The MCC assumes values in the interval [−1,1] and the MCC curve is obtained by plotting the MCC value versus the false positive rate (Fig. 2.11b).

2.5 Graph Theory

Graphs are made up of nodes which represent the neurons and edges which model the connections (morphological or functional) among the neurons. If we consider the directionality of the connection (i.e., from a pre- to a post-synaptic neuron), the graph is named directed, otherwise it is called undirected. The structure of the graph is described by the adjacency matrix (often named connectivity matrix (CM)), a square symmetric matrix of size equal to the number of nodes N with binary entries. If the element $a_{ij} = 1$, a connection between the node j to i is present, otherwise $a_{ij} = 0$ means the absence of connections. To allow a mathematical analysis, the graph, and consequently the network topology, can be characterized by a large variety of parameters [19]. In the field of neuronal networks, the simplest metrics which allow to have a simple but clear indication of the kind of underling connectivity are the

Node Degree, the Cluster Coefficient and the Average Path Length [20] which will be briefly described below.

Node Degree: the in-degree (id) and the out-degree (od) of a single node are defined as the number of incoming (afferent) and outcoming (efferent) edges respectively, and the total degree (td) is their sum.

$$td = id + od \tag{2.21}$$

High in-degree values indicate neural units influenced by a larger number of nodes, while high out-degree values show a large number of dynamic sources. Depending on the node degree distribution, we can identify three stereotyped graphs: scale-free, regular, and random (Fig. 2.12).

Scale-free networks (Fig. 2.12b, panel a) [21] are characterized by high-connected units called hubs. Hubs are nodes with a degree at least one standard deviation above

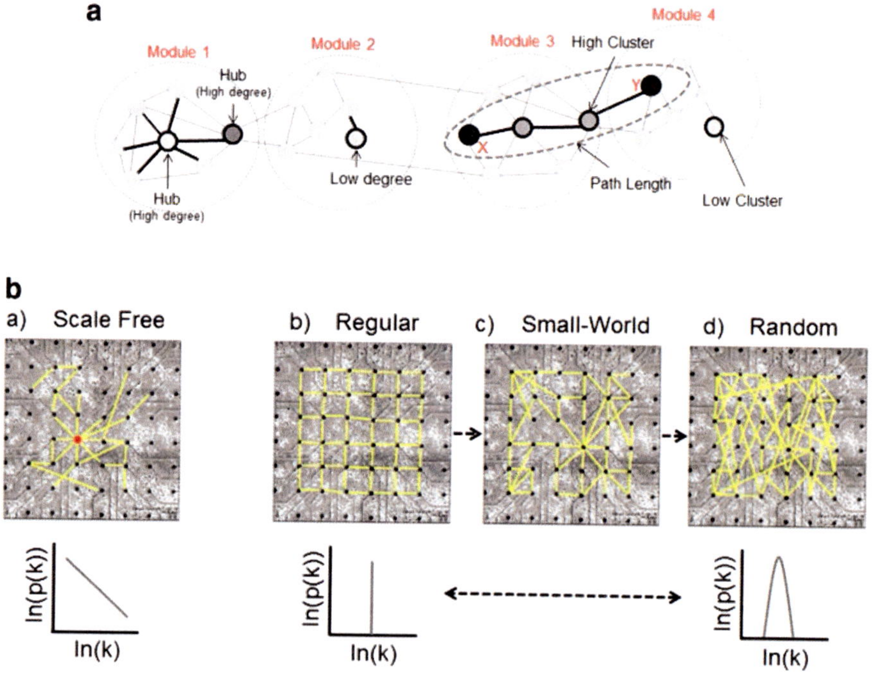

Fig. 2.12 Basic graph measures and network structures. **a** Node degree is the number of connections of a given node; this panel shows a simple network divided in four different modules: Module 1, in which we can see a high-connected unit called hub, and Module 2, that presents a low connectivity case. Modules 3 and 4 show two units with high and low values of Cluster Coefficient respectively, and an example of shortest path length; the nodes X and Y are connected by the shortest possible path (three links), and two different units that we call intermediaries. **b** Classification of the network structure [scale-free (panel a), regular (panel b), small-world (panel c) and random (panel d)] and corresponding degree distributions. Reproduced from [15]

the network mean. Thanks to this peculiarity, hubs play a significant role on the neural dynamics [22]. In the scale-free network, the probability that a generic node i has k connections is given by a power law relationship:

$$p(k) \propto k^{-\Upsilon} \tag{2.22}$$

where Υ is the characteristic exponent which ranges experimentally from 1.3 (slice recordings [23]) to 2 (fMRI recordings [24]).

Regular networks (Fig. 2.12b, panel b) are ordered and characterized by high segregation values. The integration levels of the network grow by increasing the number of graph units. In this case, the probability that i has k connections is given by:

$$p(k) = c \tag{2.23}$$

where c is a constant.

Random networks (Fig. 2.12b, panel d) show each node with a different connectivity degree and the probability that a single unit has k connections is modeled by a Poisson distribution:

$$p(k) \propto \frac{e^{-\delta}\delta^k}{k!} \tag{2.24}$$

where δ is the average connectivity degree of the network. The random graph has few local connections and therefore it shows low segregation values. The integration levels of the network, instead, follow the logarithm of the number of nodes.

A last case is the *small-world network* (Fig. 2.12b, panel c): it shares the same characteristics of regular and random networks, constituting a sort of composite model. By increasing the probability p of rewiring, the order of a regular lattice is disrupted, and when $p = 1$ a random graph is generated. Increasing the probability of rewiring, both the integration and the segregation levels decrease. In a small-world network, the distance between two nodes grows according to the logarithm of the number of nodes of the graph [25].

Cluster Coefficient (CC)

Let x be a generic node and v_x the total number of nodes adjacent to x (and including x). Let u be the total number of edges that actually exist between x and its neighbors. The maximum number of edges that can exist among all units within the neighborhood of x is given by $v_x(v_x - 1)/2$. The Cluster Coefficient (C_x) for the node x, is defined as follows

$$C_x = \frac{2 * u}{v_x(v_x - 1)} \tag{2.25}$$

The Average Cluster Coefficient, obtained by averaging the cluster coefficient of all the network's nodes, is a global metric often used to quantify the segregation at network level (Fig. 2.12a).

Average Shortest Path Length (PL)

Let x and y be two generic nodes of a network V. Let d (x, y) be the shortest distance between the nodes x and y. We define the Average Path Length (L) as follows:

$$PL = \frac{2}{n(n-1)} \sum_{x \neq y} d(x, y) \tag{2.26}$$

This topological parameter is commonly used to evaluate the network's level of integration (Fig. 2.12a).

Small World Index (SWI)

To detect the emergence of small-world network [26], it is possible to combine the metrics previously introduced, defining the Small-World Index (SWI):

$$SWI = \frac{\frac{C_{net}}{C_{rnd}}}{\frac{L_{net}}{L_{rnd}}} \tag{2.27}$$

where C_{net} and L_{net} are the cluster coefficient and the path length of the investigated network; while C_{rnd} and L_{rnd} correspond to the cluster coefficient and the path length of random networks equivalent to the original network (i.e., with the same number of nodes and links). A SWI higher than 1 suggests the emergence of a small-world topology.

Rich Club Coefficient

The Rich-Club Coefficient (RCC) [27] Φ at a specific k level is computed by evaluating the cluster coefficient between the nodes with a degree higher than k:

$$\Phi(k) = \frac{2E_{>k}}{N_{>k}(N_{>k} - 1)} \tag{2.28}$$

where $E_{>k}$ and $N_{>k}$ represent the number of edges and nodes with a degree higher than k, respectively. The RCC curve is obtained by evaluating the RCC at the varying of k from 1 to the maximum degree. The RCC is normalized by the corresponding average value for a set of surrogates random neural networks equivalent to the investigated one (i.e., networks with the same number of nodes and edges). A privileged sub-network (i.e., a rich club) emerges, if the computed normalized coefficient value is higher than one.

2.6 BIOCAM and Active Pixel Sensor (APS) Array

During my Ph.D., I studied neural cultures coupled to both low resolution and high resolution acquisition system. Briefly, the low resolution acquisition system is represented by the MEA60 and MEA2100 by Multi-Channel System (www.multichannelsystems.com; MCS, Reutlingen, Germany) 60 electrodes device, while the 3Brain APS (www.3brain.com; Landquart, Switzerland) was used as a high resolution acqusition system. In this section, I will briefly describe the BIOCAM X and the APS MEA-4 k chip. The 3Brain APS chip is a platform based on monolithic Complementary Metal Oxide Semiconductor technology enabling acquisitions from 4096 microelectrodes at a full frame rate of 9043 kHz and at a spatial resolution up to 21 μm (electrode separation) [28]. Based on the Active Pixel Sensor concept (APS), the platform development, electrical characterization and preliminary validation was performed on cardiac tissue. To introduce the platform, Fig. 2.13 presents a schematic description.

Briefly, the two essential elements are: (i) the metallic microelectrode array implemented similarly to a light imager, with in-pixel integrated microelectrodes and low-noise amplifiers; and (ii) a real-time acquisition and processing board. Based on image/video concepts implemented in hardware, the signal and data processing is furthermore adapted for handling a very large data flow (typically 0.5 Gbit/s). The platform enables acquisitions with a spatial resolution comparable to mammalian neuronal cell bodies (i.e., microelectrode size of 21 μm, electrodes separation from 21 to 81 μm), and a temporal resolution down to 8 ms/pixel on 64 selected pixelmicroelectrodes. The 3Brain system allows accessing in real-time the spatial–temporal correlation of activities at both local and network levels acquired from 4096 microelectrodes.

The BIOCAM (Fig. 2.14) is the hardware tool that allows to sample the 4096 recording electrodes simultaneously at 18 kHz, providing orders of magnitude more data points compared to conventional passive MEAs), thus leading to the recording,

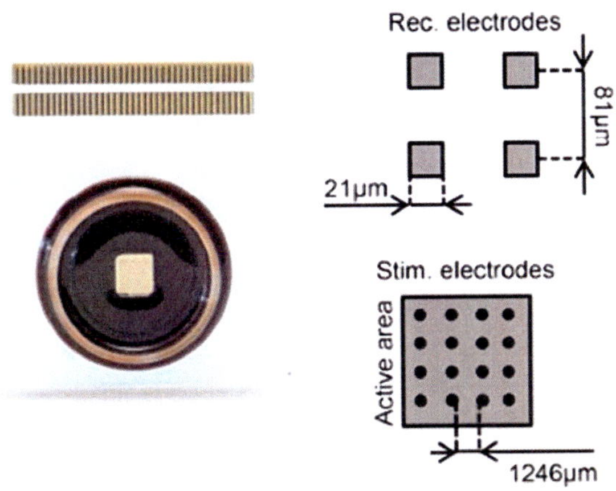

Fig. 2.13 Overview of the high resolution electrophysiological platform. The system is composed of three hardware levels, i.e. the CMOS–MEA chip, the interface board and a work station equipped with a frame grabber for capturing and storing the video stream. Adapted from www.3brain.com

Fig. 2.14 3Brain BIOCAM. Reproduced from www.3brain.com

over a large bandwidth, of electrophysiological signals ranging from slow large-population field potentials to fast single-cell spiking activity. In addition, it incorporates further optional functionalities, which come as separate modules in most MEA-systems, such as a temperature control system and an electrical programmable current-driven stimulator.

References

1. Salinas E, Sejnowski TJ (2001) Correlated neuronal activity and the flow of neural information. Nat Rev Neurosci 2:539–550
2. Eytan D, Minerbi A, Ziv N, Marom S (2004) Dopamine-induced dispersion of correlations between action potentials in networks of cortical neurons. J Neurophysiol 92:1817–1824
3. Wagenaar DA, Pine J, Potter SM (2006) An extremely rich repertoire of bursting patterns during the development of cortical cultures. BMC Neurosci 7:11
4. Pastore VP, Massobrio P, Godjoski A, Martinoia S (2018) Excitatory-inhibitory links and network topology in large scale neuronal assemblies from multi-electrode recordings. PLoS Comput Biol 14(8):e1006381. https://doi.org/10.1371/journal.pcbi.100638
5. Brillinger DR, Bryant HL, Segundo JP (1976) Identification of synaptic interactions. Biol Cybern 22:213–228. https://doi.org/10.1007/bf00365087
6. Eichler M, Dahlhaus R, Sandkuhler J (2003) Partial correlation analysis for the identification of synaptic connections. Biol Cybern 89:289–302
7. Papoulis A, Pillai SU (2002) Probability, random variables and stochastic processes, 4th edn. Mc Graw Hill, New York
8. Dahlhaus R, Eichler M, Sandkuhler J (1997) Identification of synaptic connections in neural ensembles by graphical models. J Neurosci Methods 77:93–107
9. Schreiber T (2000) Measuring information transfer. Phys Rev Lett 85:461–464
10. Bennett K, Robertson J (2011) MATLAB—A ubiquitous tool for the practical engineer. In: Ionescu CM (ed)
11. Overbey LA, Todd MD (2009) Dynamic system change detection using a modification of the transfer entropy. J Sound Vib 322:438–453. https://doi.org/10.1016/j.jsv.2008.11.025

12. Berdondini L et al (2006) A microelectrode array (MEA) integrated with clustering structures for investigating in vitro neurodynamics in confined interconnected sub-populations of neurons. Sens Actuat B Chem 114:530–541
13. Garofalo M, Nieus T, Massobrio P, Martinoia S (2009) Evaluation of the performance of information theory-based methods and cross-correlation to estimate the functional connectivity in cortical networks. PLoS ONE 4:e6482. https://doi.org/10.1371/journal.pone.0006482
14. Taketani M, Baudry M (2006) Advances in network electrophysiology: using multi-electrode arrays. Springer, New York
15. Poli D, Pastore VP, Massobrio P (2015) Functional connectivity in *in vitro* neuronal assemblies. Front Neural Circ 9:57. https://doi.org/10.3389/fncir.2015.00057
16. Nigam S et al (2016) Rich-club organization in effective connectivity among cortical neurons. J Neurosci 36:670–684. https://doi.org/10.1523/JNEUROSCI.2177-15.2016
17. Grun S, Rotter S (2010) Series in computational neuroscience. Springer, p 444
18. Fawcett T (2006) An introduction to ROC analysis. Pattern Recogn Lett 27:861–874
19. Rubinov M, Sporns O (2010) Complex network measures of brain connectivity: uses and interpretations. NeuroImage 52:1059–1069. https://doi.org/10.1016/j.neuroimage.2009.10.003
20. Sporns O, Tononi G, Edelman GM (2000) Connectivity and complexity: the relationship between neuroanatomy and brain dynamics. Neural Netw 13:909–922
21. Barabasi A-L, Bonabeau E (2003) Scale-free networks. Sci Am 50–59
22. Sporns O, Honey CJ, Kotter R (2007) Identification and classification of hubs in brain networks. PLoS ONE 2:e1049. https://doi.org/10.1371/journal.pone.0001049
23. Bonifazi P et al (2009) GABAergic hub neurons orchestrate synchrony in developing hippocampal networks. Science 326:1419–1424
24. Eguiluz VM, Chialvo DR, Cecchi GA, Baliki M, Apkarian AV (2005) Scale-free brain functional networks. Phys Rev Lett 94:018102
25. Watts DJ, Strogatz SH (1998) Collective dynamics of 'small-world' networks. Nature 393:440–442
26. Downes JH et al (2012) Emergence of a small-world functional network in cultured neurons. PLoS Comput Biol 8:e1002522. https://doi.org/10.1371/journal.pcbi.1002522
27. Schroeter MS, Charlesworth P, Kitzbichler MG, Paulsen O, Bullmore ET (2015) Emergence of rich-club topology and coordinated dynamics in development of hippocampal functional networks *in vitro*. J Neurosci 35:5459–5470. https://doi.org/10.1523/JNEUROSCI.4259-14.2015
28. Berdondini L et al (2009) Active pixel sensor array for high spatio-temporal resolution electrophysiological recordings from single cell to large scale neuronal networks. Lab Chip 9:2644–2651

Chapter 3
Results

In the first section of the Results, I will consider each of the connectivity methods that I developed and implemented during my Ph.D. In detail, I will at first show the evaluation procedure's results. Such a procedure exploits the use of an in silico neural network model made up of 1000 randomly connected neurons, characterized by an average ratio between inhibitory and excitatory connections of 1/4. Simulations display the typical signature characterized by a mix of spiking and bursting activity, as displayed by the raster plot and the Instantaneous Firing Rate (IFR) traces of the excitatory (red) and inhibitory (blue) neuronal populations of Fig. 3.6a. From a topological point of view, both the excitatory and inhibitory structural sub-networks follow a random connectivity, as the degree distributions histograms of Fig. 3.6b display. More details about the computational model can be found in the appendix of this work. The evaluation procedure assesses the computational accuracy of the connectivity methods considering them as binary classifiers, building ROC and MCC curves. After evaluating the capability to detect the in silico functional connections (e.g., looking at the ratio between true and false positive), I checked the capability of the connectivity methods to reconstruct the topology of the model (looking at topological parameters like the degree distribution, see section *Graph theory*) and the delay distribution of the in silico functional connections. At first I will report the results relative to the information-theory based methods: TE (and its extensions DTE, DHOTE), JE, and then I will describe the correlation-based ones: FNCCH and PC.

V. P. Pastore, *Estimating Functional Connectivity and Topology in Large-Scale Neuronal Assemblies*, Springer Theses, https://doi.org/10.1007/978-3-030-59042-0_3

3.1 Transfer Entropy

3.1.1 Validation of the TE by Means of in Silico Neural Networks

I tested the developed DTE and DHOTE algorithms on 5 realizations of the in silico neuronal network model described in the appendix. I randomly extracted 500 neurons from such simulated neuronal populations, and I run the TE, the DTE and DHOTE algorithms over one hour of simulation (sampled at 1 kHz).

First of all, I tested and evaluated our classic TE algorithm. To determine the optimal bin value that maximizes the accuracy, I considered the TPR value corresponding to a FPR = 0.01. Figure 3.1a shows the obtained results. The accuracy displays a sharp increasing trend, reaching a maximum for a bin of 9 ms. Then, in order to compare classic TE, DTE and DHOTE, I evaluated the accuracy of the methods by means of the Receiver Operating Characteristic (ROC) curves [1]. Within this framework, the True Positive Rate (TPR) and the False Positive Rate (FPR) are defined on the basis of the true positive (TP), true negative (TN), false positive (FP) and false negative (FN) values. First of all, I explored the variation of the accuracy with respect to the order k and l for DTE. I considered a temporal window of 30 ms for computing the DTE, corresponding to a classic TE (with bin equal to 1 ms) computed for 30 different delays, as explained in section *Delayed Transfer Entropy*. Following the approach used in [2], I took into account the TPR value corresponding to a very low value of the FPR (0.01) performing the evaluation by varying the parameters from order (1, 1) to order (3, 3). Figure 3.1b, c show in terms of false color maps the results obtained considering the peak value (Fig. 3.1b) and maximum CI (Fig. 3.1c) of DHOTE. The best accuracy performances in both cases are achieved by using k = 1 and l = 3 (red asterisk). As expected, the computation time (Fig. 3.1d) increases with the total order (k, l) although it remains reasonably low for all the explored orders. For 60 min of simulation of 500 neurons, the maximum computation time is 27 min (54 s for a single TE computation) for order (3, 3). The computation time required for the order (1, 1) is equal to 9 min (18 s for a single TE computation). Finally, I compared the TE (using the optimum bin of 9 ms) with the DTE and the DHOTE with k = 1 and l = 3 by means of the ROC curves. Figure 3.1e shows the obtained results, while in Fig. 3.1f we can see the corresponding Area Under Curve (AUC) values. There is a relevant improvement in the accuracy consequent to the temporal extension of the TE method, while the extension of the TE's order to k = 1 and l = 3 produces only little improvement with respect to TE (Fig. 3.1f). In the light of the previous analysis and considerations, I used the DHOTE showing the best performances (k = 1 and l = 3), to explore the capability of such a method to reconstruct the temporal delay and the degree distribution of the in silico neuronal network. Therefore, a hard threshold procedure to extract the strongest effective links from the TE's connectivity matrix was used. In particular, following the results achieved in [3], I used a threshold equal to $\mu + \sigma$, where μ and σ are the mean and the standard deviation values of the TE's connectivity matrix, respectively. Figure 3.1g,

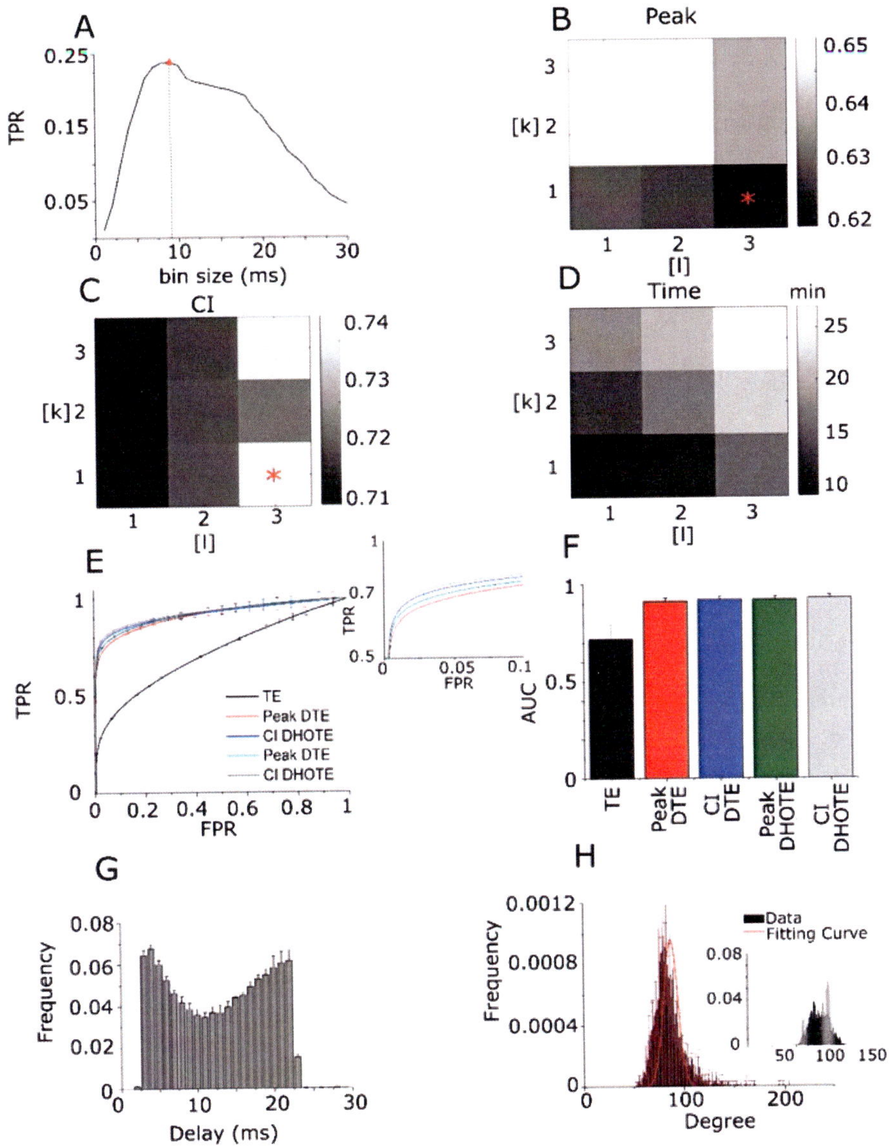

Fig. 3.1 TE's testing on in silico neural networks. **a** TPR (corresponding to FPR = 0.01) for classic TE computed by increasing the bin size from 1 to 30 ms. **b**, **c**, TPR corresponding to FPR = 0.01 for increasing TE's orders for reference train (k) and for target train (l) is considered. The red symbol "*" indicates the (k, l) pair that guarantees the best accuracy. (B) Values computed using TE's peak. **c** Values computed using TE's CI. **d** Computational times for increasing TE's orders. **e** ROC curves corresponding to TE (black), DTE (peak and CI, red and green, respectively) and DHOTE (peak and CI, blue and gray, respectively); inset of panel (**e**) shows a zoom of the same curve. **f** AUC derived from the ROC curves of panel e. **g** Delay distribution obtained using the DHOTE ($k = 1$, $l = 3$) CI. The reconstructed delay distribution is coherent to the one used to generate the neural model. **h** Degree distribution obtained using the DHOTE ($k = 1, l = 3$) CI. The reconstructed degree distribution fits a Gaussian curve that is the distibution used to generate the underlying connectivity of the in silico neuronal model (inset). Reproduced from [4]

h display how DHOTE reconstructs both delay and degree distribution of the simulated network. Figure 3.1g shows that the reconstructed delay distribution from the functional-effective links is qualitatively similar to the uniform distribution of the delays of the neuronal network model. Figure 3.1h shows that the estimated degree distribution can be fitted with a Gaussian distribution ($R^2 = 0.89$), in accordance with the original distribution of the structural connectivity of the in silico neuronal network (Fig. 3.1h inset).

3.1.2 DTE Application to Neural Networks Coupled to the MEA-4k

In this section, I report a demonstrative example of the use of DTE to process and analyze 30 min of spontaneous activity of a cortical network recorded by means of a 4096 HD-MEA (3Brain) at the sampling frequency of 9043 Hz. I computed the TE on the extracted spike trains in order to characterize the topological features of the network connectivity. I used a first order ($k = 1, l = 1$) DTE with a temporal window of 20 ms and with a bin size equal to the sampling period (0.12 ms). Thus, to cover the 20 ms temporal window I computed 170 TE corresponding to 170 different delays. Figure 3.2e shows the connectivity graph corresponding to the DTE analysis of an illustrative cortical network. To provide some quantitative results, I performed some topological analysis (Fig. 3.2f) by evaluating the CC, PL, SWI, and percentage of hubs. The SWI is equal to 4.8 suggesting that the considered network exhibits small world features; the degree distribution is plotted in a log–log plane in Fig. 3.2g.

3.2 Joint Entropy

3.2.1 Shuffling Thresholding Procedure Validation on in Silico Networks

The thresholding approach based on surrogate data described in the section *Shuffling thresholding procedure*, requires the choice of three main parameters. The first one is the number of surrogates to generate for the statistical significance test (i.e., the number of times the electrodes' time stamps will be randomly displaced and shuffled), while the second one is the parameter that I called *delta*, used to compute the minimum difference between shuffled JE and original JE value to take into account for rejecting the null hypothesis, considering a functional connection significant (see section *Shuffling approach*). In detail, the minimum difference will be defined as delta multiplied by the average value of a difference distribution built with a bootstrap shuffling approach (see, Shuffling thresholding procedure). The last parameter is the p-value, that represents the number of times over the total allowing the null

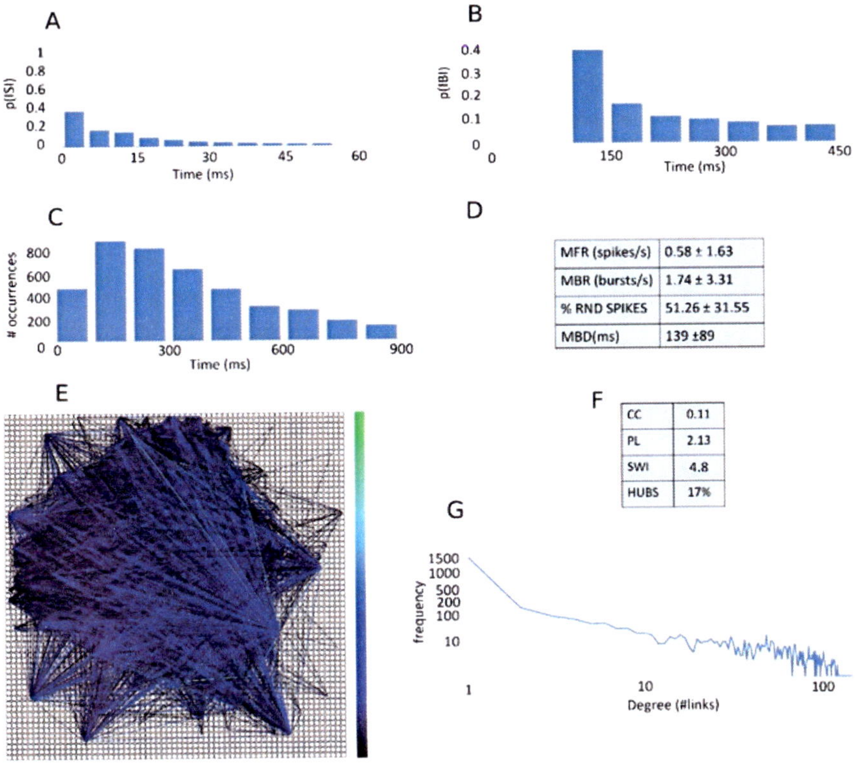

Fig. 3.2 Dynamic, connectivity and topological analysis of a cortical neural network coupled to the 4096 electrodes 3Brain system. **a** ISI histogram (bin = 5 ms). **b** IBI histogram (bin = 50 ms). **c** Burst duration histogram (bin = 100 ms). **d** Summary of the spiking and bursting statistics, **e** Connectivity graph obtained by means of DTE thresholded with $\mu + 4\sigma$ to show only the strongest links. **f** Summary of the topological parameters. **g** Degree distribution. Reproduced from [4]

hypothesis to be rejected before considering as not significant an investigated connection. For instance, a p-value of 0.01 indicates that if the original JE value is higher than the shuffled one, more than 1 time over 100, I will discard the correspondent link considering it as not significant. To validate the method, and tuning plausible values for such parameters, I applied the JE algorithm to 10 realizations of the in silico neural networks, described in detail in the appendix. I tested the shuffling approach using 100 and 1000 surrogates for each neuron, sweeping the parameter delta from 0 to 3 with unitary step. I evaluated the shuffling performances in terms of precision, evaluating the ratio between TP and FP. In detail, the precision is defined as TP/(TP + FP). The higher the precision, the higher the performances of a binary classifier. In our case, the binary classifier is based on the JE algorithm, with different performances in function of the parameter *delta*, the p-value and the number of surrogates used. At first, I tested the classical shuffling approach, in which both the reference and the target train are managed with random spike displacements to destroy an eventual correlation between spike trains building a null hypothesis correspondent to non-connections. Using 100 surrogates, I obtained a precision value of 0.15 ± 0.04. Then, I tested our shuffling approach, in which only the target train is shuffled, to be more restrictive obtaining a lower number of functional links with a higher precision. Figure 3.3a–d show the average results obtained for 100 and 1000 surrogates. The 3-D graph correlates precision with the parameter delta and the p-value used for the significance statistical test. As we can see, a delta value of 3 with a p-value of 0.01 corresponds to a maximum value of precision for both the number of surrogates. The generation of the surrogate data, and above all, the statistical tests, need the execution of the JE algorithm for each couple reference electrode-shuffled target electrode, thus require a computational time that is function of the number of surrogates. According to this, considering that the developed algorithm has to be used to analyze neural networks coupled to the HD-MEA acquisition system, with 4096 electrodes, the best choice seems to be the use of 100 surrogates with a p-value of 0.01 and a *delta* value of 3. I use such parameters to analyze two cortex neural networks coupled to the HD-MEA system (JE Application to cortex network coupled to the HD-MEA).

3.2.2 Hard Threshold Validation on in Silico Neural Networks

The shuffling approach in thresholding the connectivity matrix is aimed to provide an efficient method, with statistical basis, to select the meaningful functional connectivity links. However, the simplest approach remains the use of a hard threshold procedure as described in *Shuffling thresholding procedure* section. The main problem with such a thresholding approach, is the choice of the parameter n. As Fig. 3.3e shows, the precision critically depends on the choice of n. In particular, the precision increases at the increasing of n, with a consequent decreasing of the number of true positive links. The hard threshold approach is based on the assumption that the strongest the links (i.e., the lowest the obtained value for the JE), the most likely a found connection can be an actual one. To support such a hypothesis, I considered the JE values correspondent to the functional links that survived the shuffling thresholding procedure. As Fig. 3.3f shows, the JE values that survived to the shuffling thresholding procedure, correspond to the lowest ones in the CM, and would have been selected by the hard thresholding procedure, being a subset of the links surviving such a procedure (red line n = 3, blue line n = 4). However, it is worth noticing that the shuffling approach is independent from the user in the thresholding procedure. The AUC value is equal to 0.6848 ± 0.033, that is higher than 0.5 (value correspondent to random guess), but not descriptive of good classification performances. In fact, when applying a threshold to select the most significant functional connections, we work in an area of FPR rate that are very low. Thus, the TPR rate should be high in this region to be representative of good classification performances. The CC method

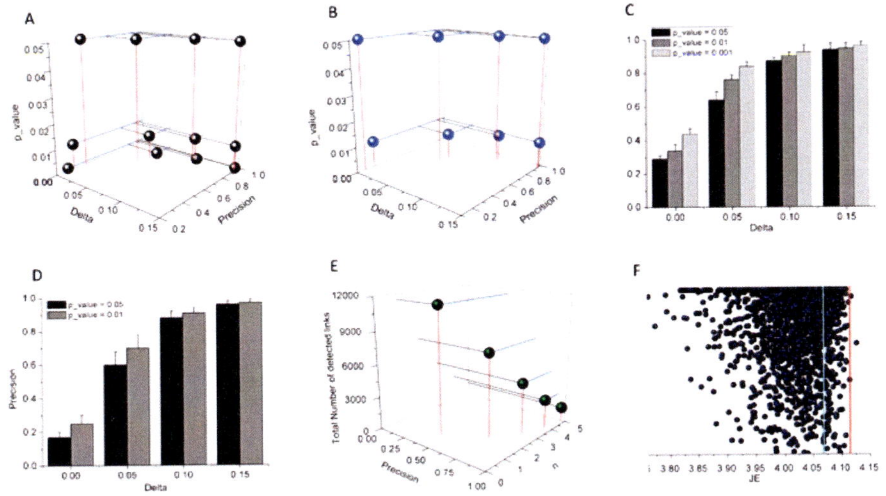

Fig. 3.3 Shuffling thresholding procedure validation on in silico neural networks. **a**, **b** Average values of precision for the shuffling approach at the varying of the parameters p_value and Delta in a 3d graph for a number of surrogates = 1000 (**a**) and 100 (**b**). **c**, **d** Same information in a 2d graph for a number of surrogates = 1000 (**c**) and 100 (**d**). **e** Average values of the precision and the total number of detected links for the hard thresholding approach at the varying of n. **f** JE values corresponding to the links surviving the shuffling thresholding procedure. These links are a subset of the hard thresholding procedure survival links (with n = 3, red line and n = 4, blue line)

(blue curve) shows a behavior similar to the DTE one, correspondent to poor performances in detecting and identifying the model functional connections. The MCC curves (Fig. 3.4g) confirm the information provided by the ROC curves. In fact, the JE (red curve) presents its maximum value (0.601 ± 0.005) in correspondence to a FPR value of 0.01, indicating good classification performances. Both the DTE (black curve) and the CCH (blue curve), instead, show negative values in correspondence of FPR value of 0.01, representing a complete disagreement between prediction and observation. The reason of such impressive differences between JE, CC and DTE has to be searched in the inhibitory links detection and identification. In fact, it is worth to notice that JE is not able, by definition, nor to detect or to identify inhibitory links. Thus, the curve showed in Fig. 3.4c, actually, is relative only to the excitatory links. The DTE, as well as the CC, are instead able to detect, by definition, inhibitory links, but they are not able to identify it, that is, they are not able to distinguish excitatory links from inhibitory links. It is worth to notice, that inhibitory neurons have a firing rate higher than the excitatory ones (5–6 spikes/s versus 2 spikes/s in the model). According to this, both the DTE and the CC detect strong functional links correspondent to inhibitory links, due to artifacts and not to actual connections. To prove such statement, that is to show that DTE and CC are hardly influenced and wasted by the inhibitory neurons and links, I considered for the ROC curve analysis only the excitatory neurons (the first 800). Figure 3.4d, h show the correspondent ROC curve and MCC curves. As we can see, the DTE and the CC hardly improve their

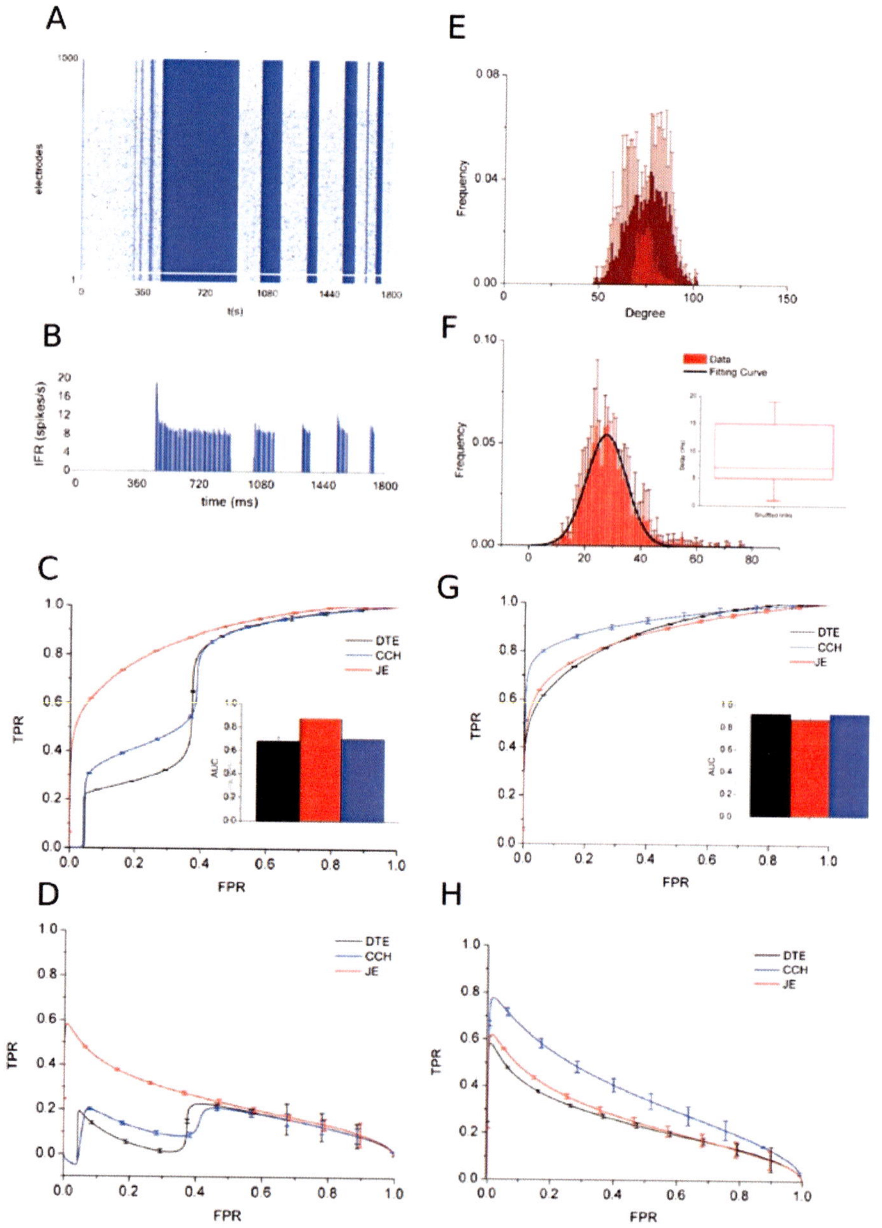

Fig. 3.4 JE validation on in silico neural networks. **a** Raster Plot and **b** mean Instantaneous Firing Rate (IFR) representative of the simulated spiking activity. **c** ROC curves and MCC curves (**d**) relative to DTE (black curve), JE (red curve) and CCH (blue curve) considering both the excitatory and the inhibitory neurons. Correspondent AUCs in the inset. When only the excitatory sub-networks are considered, DTE and CCH severally improve their performances, as shown by the ROC curves (**g**) and MCC curves (**h**). Using the shuffling approach, the degree (**f**) and the delay distribution (**f**, inset) is reconstructed. Such distributions reflect the ones used to generate the in-silico networks model (**e**)

performances, while the JE's curve is almost the same. In the excitatory subnetwork, the CC algorithm shows the best performances reaching a TPR value of 0.72 ± 0.01 in correspondence to the false positive rate of 0.01. The MCC curve confirms what emerges from the ROC analysis, with the CC reaching a peak of 0.77 ± 0.02. The DTE shows a similar behavior with a TPR value of 0.50 ± 0.01 in correspondence of a FPR value of 0.01, and a MCC peak equal to 0.57 ± 0.01. The JE algorithm, instead, has a TPR value of 0.580 ± 0.005 in correspondence of a FPR value of 0.01, and a MCC peak equal to 0.616 ± 0.004. Later, I applied the shuffling approach with 100 surrogates and a delta value of 3 (see Shuffling validation on in silico networks) to the JE connectivity matrices. I used the thresholded connectivity matrices to extract the mean degree distribution (Fig. 3.4f). The computed average distribution fits a Gaussian Curve ($R^2 = 0.92$, black curve), that is the distribution used to generate the model (Fig. 3.4e). Finally, I computed the delay distribution for the excitatory links (Fig. 3.4f, inset). The obtained distribution reflects the uniform distribution in [0, 20] ms used to generate the in silico model.

3.2.3 JE Application to Cortex Network Coupled to the HD-MEA

Finally, I applied the developed JE algorithm to two cortical neural networks coupled to the HD-MEA acquisition system, after they reached a stable stage (i.e., after 21 days in vitro). In particular, I analyzed two networks characterized by two different seeding densities: a low density network (400 cell/mm^2) and a high density one (1000 cell/mm^2). It is possible to see an example of the recorded spontaneous electrophysiological activity in form of raster plot (Fig. 3.5a) and instantaneous firing rate (Fig. 3.5b). I used our developed JE algorithm with the implemented customized shuffling approach to infer the functional connectivity of the two neural networks. Figure 3.5c represents an example of the thresholded connectivity graph (using the shuffling thresholding approach with 100 surrogates, delta = 3 and a p-value of 0.01), where nodes and edges represent electrodes and functional links, respectively. I performed a topological analysis (Fig. 3.5e) extracting the cluster coefficient and the average path length [5]. The low density neural network showed a higher cluster coefficient than the high density one (0.192 vs. 0.042). Such a difference is likely the reflection of the cross-talking effect with high seeding densities in the in vitro model. I extracted the Small World Index (SWI) by comparing the cluster coefficient and the average path length of our networks with the average values extracted from 100 random networks equivalent to the investigated one (i.e., with the same number of nodes and links, as done in [6]). Both the networks showed the emergence of a small world topology (Fig. 3.5e). Such a topology is coherent with the degree distribution, represented in Fig. 3.5d. Again, the difference in the absolute value of the SWI (10.57 vs. 2.15) is likely referred to the different cellular seeding densities.

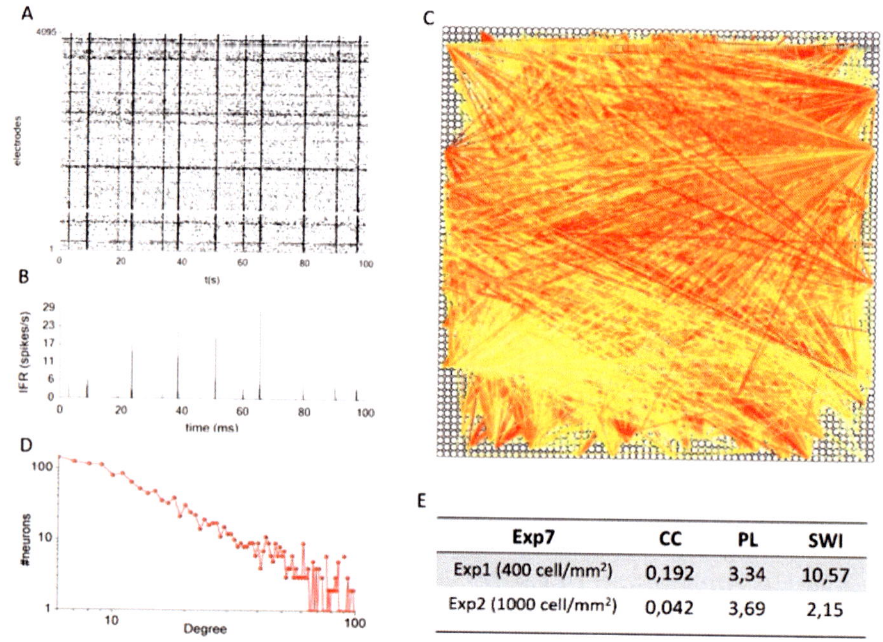

Exp7	CC	PL	SWI
Exp1 (400 cell/mm²)	0,192	3,34	10,57
Exp2 (1000 cell/mm²)	0,042	3,69	2,15

Fig. 3.5 JE application to cortex networks coupled to the HD-MEA acquisition system. **a** Raster Plot and **b** mean Instantaneous Firing Rate (IFR) representative of the recorded spiking activity. **c** Connectivity graph in false colors and correspondent degree distribution (**d**) obtained by means of the JE algorithm and the customized shuffling approach. **e** Table reporting the main topological parameters extracted for the two networks with different seeding densities (400 cell/mm^2 and 1000 mm^2)

3.2.3.1 JE Results Discussion

The JE algorithm is characterized by both high computational accuracy and efficiency. The computational efficiency is a result of the implementation I did, that adapted the algorithm to be directly applied on the spike time stamps. The computational accuracy is linked to the selectivity of the JE algorithm for the excitatory links. In fact, JE is not able, by definition, to neither detect nor to identify the inhibitory links. The inhibitory neurons are usually characterized by high (and likely tonic) firing rate, and as I showed in section *JE Validation on* in silico*neural networks*, they can be responsible for the appearance of false positive artifacts when using other connectivity methods like DTE and CC. Moreover, the incapability to detect inhibitory links could be very useful when combining the JE algorithm to another one able to detect and identify inhibitory links. In this way, the JE could be used as a sort of inhibition filter to improve the classification performances. Thresholding the connectivity matrix is a main problem in the functional connectivity analysis, and can dramatically alter the results influencing even the kind of topology that is detected [7]. The hard thresholding approach is hardly dependent on the choice of the threshold value as I showed in section *Hard threshold validation on* in silico*neural networks*. To overcome this problem, I developed a customized version of the shuffling approach. The basic thresholding approach corresponded to low values of precision, due to the presence of a high number of false positives; thus, I introduced another threshold that is dependent on the shuffling values distribution, and so it is dependent on the network firing rate and dynamic parameters. Such thresholding approach brought to very high level of precision when tested on in silico networks.

3.3 FNCCH Results

3.3.1 Validation of the FNCCH by Means of in Silico Neural Networks

I applied the FNCCH to 10 realizations of the in silico neural networks. Figure 3.6c shows the ROC curves obtained by comparing the Synaptic Weight Matrix (SWM) of the model (i.e., the ground truth) with the computed Functional Connectivity Matrix (FCM). Figure 3.6d shows the MCC curve (cf., Methods). The ROC curve relative to the detection of inhibitory connections (blue curve in Fig. 3.6c) is very close to the perfect classifier, with an Area Under Curve (AUC) of 0.98 ± 0.01(blue bar in the inset of Fig. 3.6c). The MCC curve relative to the inhibitory links (blue curve in Fig. 3.6d) has a maximum value of 0.87 ± 0.04. Then, I compared the sensitivity of the FNCCH for the detection of excitatory links (red curves in Fig. 3.6c, d) to the standard CCH's one (for excitation, black curves in Fig. 3.6c, d), to underline the improved detection capabilities with the filtering procedure. I observed not only a significant ($p < 0.001$) AUC increase (0.92 ± 0.01 vs. 0.72 ± 0.02, Fig. 3.6c inset), but also significant improvements in both ROC and MCC curves shape for low values of false positive rates (FPR). In particular, we can notice (Fig. 3.6d), that the FNCCH excitatory curve has a maximum value of about 0.75 with respect to the correspondent NCCH's value (for the same false positive rate) that is negative (suggesting a disagreement between prediction and observation). The above results justify the use of a hard threshold procedure (cf., Methods) to select the strongest and significant functional connections. The Thresholded Connectivity Matrix (TCM) is thus directly computed from the FCM using a threshold equal to $\mu + 1 \cdot \sigma$, (mean plus one standard deviation of the connections strength) for the inhibitory links, and $\mu + 2 \cdot \sigma$ for the excitatory ones, obtaining estimated links with a very high level of accuracy (cf. Methods): $R^2 = 0.99$ for the inhibitory links and $R^2 = 0.94$ for the excitatory ones. To investigate whether the reconstructed functional connectivity network resembles the one of the model, I calculated the excitatory and the inhibitory (Fig. 3.6e) links' degree distribution from the TCM. The computed degree distributions fit a Gaussian distribution (Fig. 3.6e, $R^2 = 0.99$ for the inhibitory links and $R^2 = 0.98$ for the excitatory ones), in accordance to the original distributions used to generate the structural (random) connectivity of the model (Fig. 3.6b). It can be noticed, that

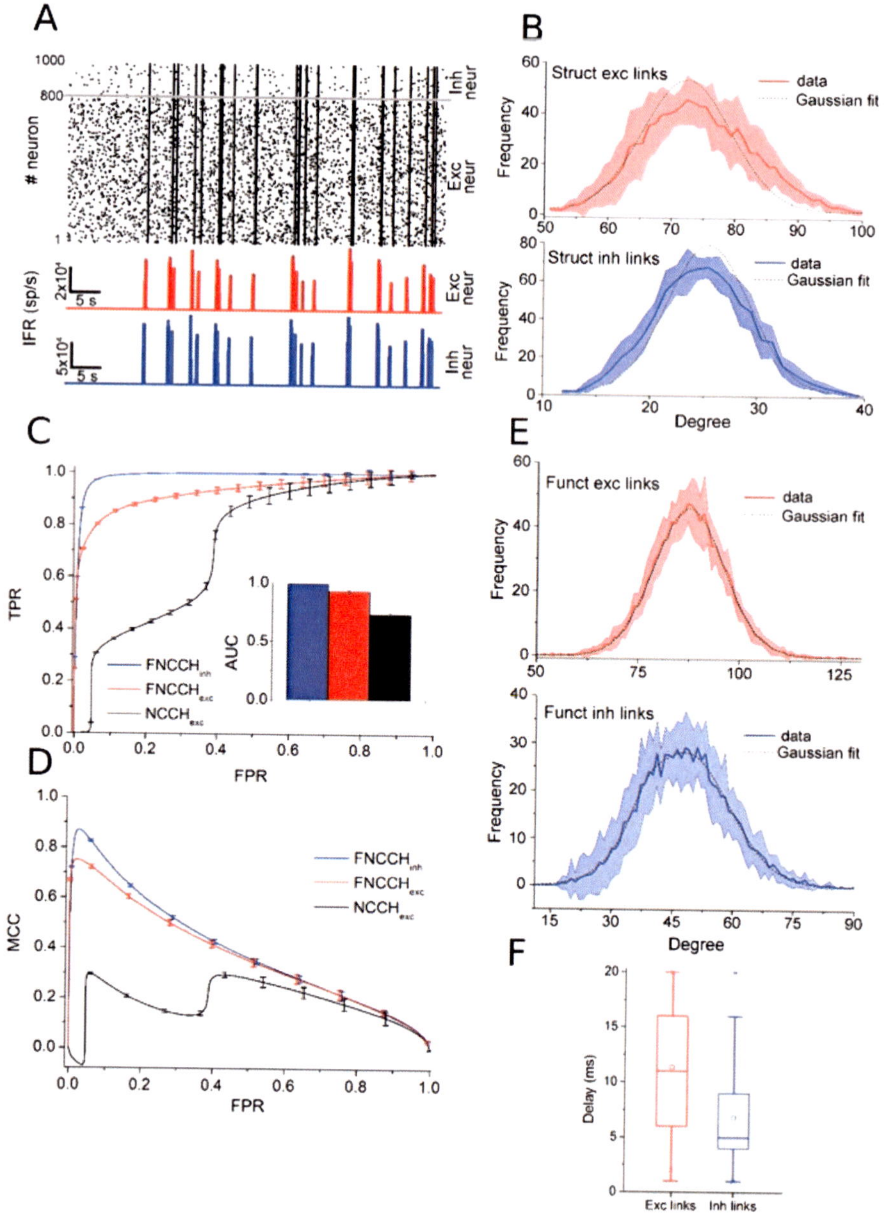
A
B
C
D
E
F
neuron
Inh neur
Exc neur
IFR (sp/s)
Struct exc links
Struct inh links
Funct exc links
Funct inh links
data
Gaussian fit
Frequency
Degree
TPR
FPR
AUC
MCC
FNCCH_inh
FNCCH_exc
NCCH_exc
Delay (ms)
Exc links
Inh links

◂**Fig. 3.6** Effective connectivity estimation from 10 in silico neural networks. **a** Raster Plot and mean Instantaneous Firing Rate (IFR) representative of the simulated spiking activity. **b** Structural degree distribution of the model (red curve for excitatory links and blue curve for the inhibitory ones). **c** ROC curves relative to inhibitory (blue curve) and excitatory (red curve) links computed by applying the FNCCH; the black curve, related to only excitatory links extracted with the standard NCCH, has been superimposed for comparison. Corresponding AUCs are represented in the inset. **d** MCC curves related to inhibitory and excitatory links computed by applying the FNCCH; the black curve, related to only excitatory links extracted with the standard NCCH, has been superimposed for comparison. **e** functional degree distribution reconstructed by means of FNCCH. The reconstructed degree distributions fit the Gaussian distribution used to generate the structural connectivity of the in silico model. **f** Box plot of the excitatory and inhibitory delay distribution obtained by means of the FNCCH. Adapted from [9]

the mean values of the functional Gaussian distributions are slightly higher than the structural ones due to the presence of false positives. Finally, I computed the delay distribution for both the excitatory and the inhibitory links of the TCM (Fig. 3.6f); both the extracted distributions reflected the ones used to generate the model. In fact, the excitatory delay distribution was uniform in the interval [0, 20] ms, while all the inhibitory links introduce a constant delay set at 5 ms (cf., Methods).

3.3.2 *Functional Connectivity and Emergent Network Topologies in In Vitro Large-Scale Neural Networks*

The FNCCH was applied to neuronal networks coupled to two different devices: MEA-60 and MEA-4k. Figure 3.7 shows the two utilized micro transducers (Fig. 3.7a, d) and the relative images of representative networks coupled to the two devices (Fig. 3.7b, e). Such networks are the morphological substrate originating the complex electrophysiological activity characterized by an extensive bursting dynamics (i.e., highly synchronized network bursts) and random spiking activity. Figure 3.7c, f show two examples of spontaneous activity recorded by a MEA-60 (Fig. 3.7c) and a MEA-4k (Fig. 3.7f). We can observe silent periods, (where very few activity appears), desynchronized spiking activity and huge bumps of activity (of different duration) called network bursts which cause a fast increasing of the firing rate as the mean Instantaneous Firing Rate (IFR) plots show (Fig. 3.7c, f). I analyzed three cortical and three striatal networks coupled to the MEA-60 (FNCCH parameters: time window $W = 25$ ms and time bin of 0.1 ms) and three cortical networks coupled to the MEA-4k (FNCCH parameters: time windows $W = 24$ ms and time bin of 0.12 ms) after they reached a stable stage (i.e., after 21 Days In Vitro, 21 DIV).

Figure 3.8a, g show a directed graph relative to a cortical and a striatal network coupled to a MEA60 device (Fig. 3.8b, c, h and i show the contribution of excitation and inhibition respectively). All the graphs were obtained by applying the hard threshold approach and a spatio-temporal filtering to prune co-activations (cf.

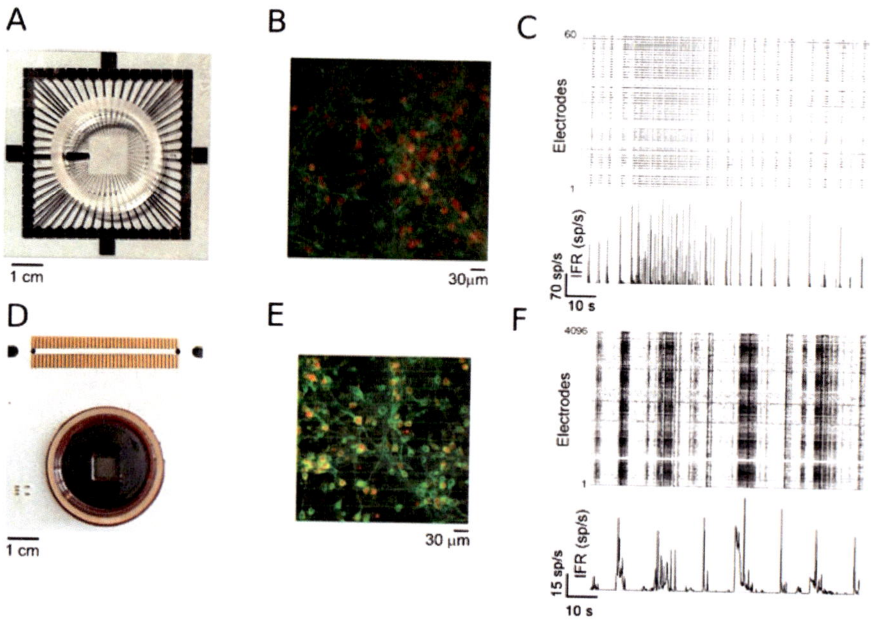

Fig. 3.7 Micro-Electrode Arrays (MEAs) used in the presented experiments. **a** MEA-60 device, **b** Cortical network coupled to a MEA-60. **c** Example of 100 s recording of spontaneous electrophysiological activity and mean Instantaneous Firing Rate (IFR) plot. **d** MEA-4k device, **e** cortical network coupled to the MEA-4k. **f** Example of 100 s recording of spontaneous electrophysiological activity and mean IFR plot. Both the recordings come from cortical assemblies at DIV 25. Reproduced from [9]

Methods). For the striatal culture, the qualitative prevalence of inhibitory connections is clearly visible. To characterize the detected links for the cortical cultures, I computed the box plots of the functional connections' delay (Fig. 3.8d) and connection length (Fig. 3.8e) of excitatory (red) and inhibitory (blue) connections. Interestingly, I found that the inhibitory links are slower and with longer connections than the excitatory ones, as previously reported in brain slices [8] for structural and functional connectivity. Similar representative graphs are shown in Fig. 3.9, where a directed graph relative to a cortical network coupled to a MEA-4k is presented (Fig. 3.9a). Link strengths are represented by two distinct color codes (arbitrary unit) to facilitate the observation of excitation (hot-red color code) and inhibition (cold-blue color code). The two detected sub-networks are also shown in Fig. 3.9b c. Moreover, box plots showing the connectivity peak delays and lengths are presented in Fig. 3.9d, e. The obtained results are in agreement with what obtained with MEA-60 devices.

I also computed the inhibitory links percentage with respect to the total number of detected links for the three different experimental conditions (n = 3 experiments for each condition). As expected, I found that striatal cultures have a higher percentage of inhibition and inhibitory links (about 60%) [10, 11] than cortical ones (about 25%). It is worth noticing that for the cortical cultures the excitatory/inhibitory ratio is

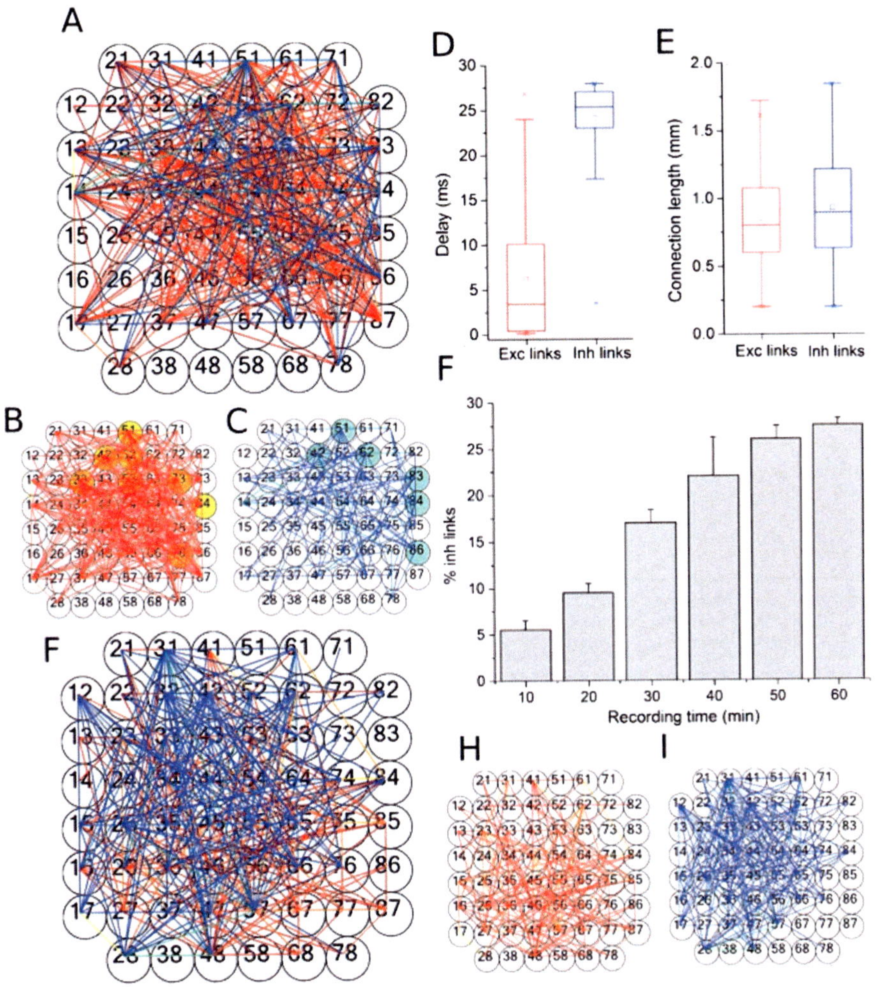

Fig. 3.8 Functional Connectivity analysis on different neural network populations coupled to MEA-60 device. **a** Functional connectivity graph obtained by applying the FNCCH to a cortical network at DIV 25. Excitatory and inhibitory links are separately thresholded and shown for convenience in panel **b** (excitation, red color map) and **c** (inhibition, blue color map). Color scales are indicative of the relative connection's strength based on the peak of FNCCH. Yellow circles in panel b and cyan circles in panel c represent the identified rich club nodes. **d** Box plot of the delays of the detected functional links. **e** Box plot of the connection lengths of the detected links. **f** Mean percentage of the inhibitory links revealed by the FNCCH at the varying of the recording time length. **g** Example of functional connectivity graph relative to a striatal network at DIV 21 coupled to a MEA-60 device. Panels **h**, **i** show the excitatory (red color map) and inhibitory (blue color map) networks, respectively. Reproduced from [9]

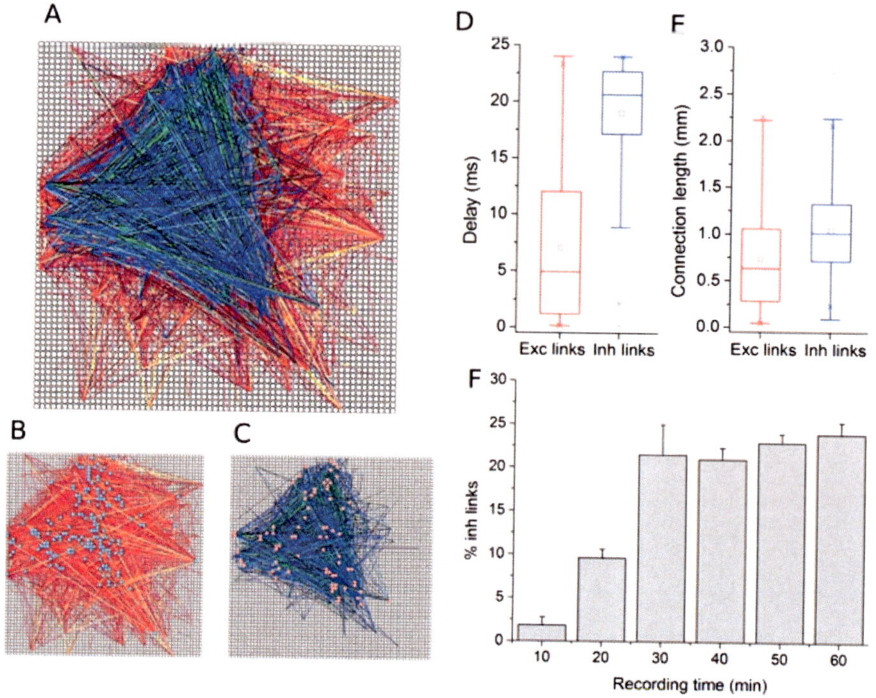

Fig. 3.9 Functional Connectivity analysis on cortex neural networks populations coupled to the MEA-4k. **a** Functional connectivity graph obtained by applying the FNCCH to a cortical network at DIV 21 coupled to a MEA-4k device. Excitatory and inhibitory links are separately thresholded and shown for convenience in panel **b** (excitation, red color map) and **c** (inhibition, blue color map). Color scales are indicative of the relative connection's strength based on the peak of FNCCH. Cyan circles in panel (**b**), and pink circles in panel (**c**), represent the identified rich club sub-networks. **d** Box plot of statistical distribution of the delays of the detected functional links. **e** Box plot of the statistical distribution of the connection lengths of the detected links. **f** Percentage of the inhibitory links revealed by the FNCCH at the varying of the recording time length. Adapted from [9]

detected quite independently from the number of recording sites (Fig. 3.8f) although it tends to stabilize with a shorter recording time for the MEA-4k. Interestingly, the found ratio, in cortical networks, between inhibitory and excitatory links (about 1/4) is about the same as the ratio of inhibitory and excitatory neurons estimated by immunostaining [12].

In order to derive the topological features [13] of the analyzed cortical networks, I computed the Clustering Coefficient, CC (Fig. 3.10a) and the average shortest Path Length, PL (Fig. 3.10b). Then, I extracted the Small-World Index (SWI) by comparing the CC and the PL of the analyzed networks to the mean values of CC and PL of 100 realizations of a random network with the same degree-distribution, as recently proposed [2]. I found that when the cortical networks are coupled to MEA-4k devices we could see the emergence of a clear small-world (SW) topology (Fig. 3.10c), while for cortical networks coupled to MEA-60 s we could not infer

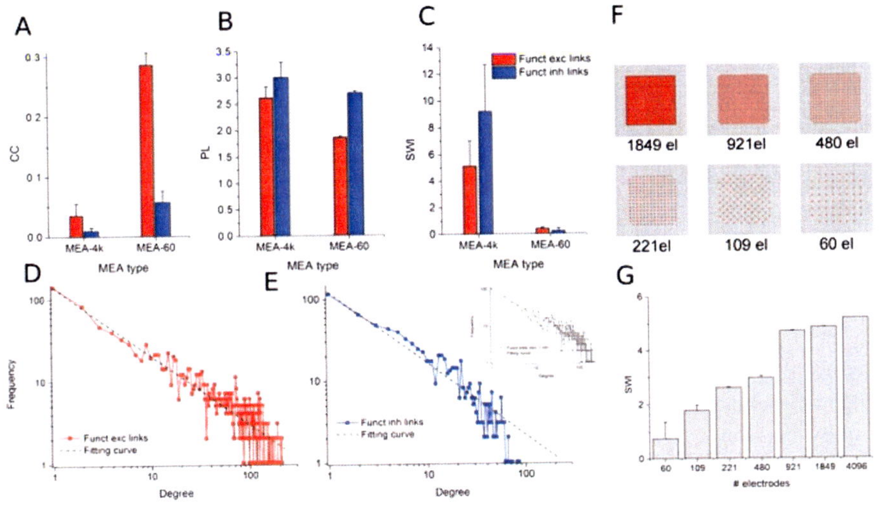

Fig. 3.10 Topological features of the detected functional networks. **a** Mean Cluster Coefficient (CC). **b** Average shortest Path Length (PL). **c** Small-World Index (SWI). Red and blue colors indicate excitatory and inhibitory population, respectively. Degree Distributions of **d** excitatory, **e** inhibitory, and total links (inset). **f** Schematic representation of the procedure used to decrease the electrodes density to analyze the SWI dependence on the electrodes' resolution. **g** SWI evaluation as a function of the electrodes density from 60 to 4096 microelectrodes. Reproduced from [9]

any SW topology. With the measurements performed with MEA-4k's I can state that, both inhibitory and excitatory links with their small world index, SWI \gg 1 (9.2 ± 3.5 for the inhibitory links and 5.2 ± 2 for the excitatory ones) contribute to 'segregation' with the emergence of small world networks. Moreover, inhibitory links, with their somehow longer connections, might contribute also to network 'integration' (i.e., communication among the SWs). To further characterize the topology of these neuronal assemblies, I also investigated the possible emergence of scale-free topologies [14] by evaluating the presence of hubs [15] and power laws for the excitatory (Fig. 3.10d), inhibitory (Fig. 3.10e) and global (Fig. 3.10e, inset) link degree distributions. In agreement with what was obtained in previously published model systems [16] and in other studies [8], I obtained that such distributions fit a power law with a R^2 higher than 0.92, in all the three cases. Finally, I searched for the presence of privileged sub-networks, constituted by the most connected nodes (i.e., rich club) of the investigated networks [17]. I identified the Rich Club Coefficient (RCC). For the analyzed cortical cultures, I found privileged sub-networks as indicated by the computed RCC with a max value of 2.7 ± 0.5. Figure 3.9b, c show the rich club networks identified for one neural network coupled to the MEA-4k, represented by means of blue circles (for excitatory subnetwork) and pink circles (for inhibitory subnetwork). Figure 3.8b, c are the analogous for a cortical neural network coupled the MEA-60 (yellow for the excitatory nodes and light blue for the inhibitory ones).

When analyzing similar cortical networks but coupled to the MEA-60 devices, as pointed out above, no SW topology is identified (Fig. 3.10c); they seem to be characterized by a sub-random topology with a SWI of 0.4 ± 0.1 for the excitatory and 0.2 ± 0.2 for the inhibitory links. These cortical networks are of the same type as the ones coupled to the MEA-4k (i.e., similar density of neurons, same age, same culture medium) and the apparent estimated random topology should be attributed to the low number of recording sites (i.e., 60 channels) that are not enough to reliably infer topological features. For determining how number and density of electrodes are crucial, I computed the SWI by considering a reduced number of electrodes for the functional connectivity analysis from the MEA-4k recording, as described in Fig. 3.10f. In particular, I started from the full resolution of the MEA-4k (i.e., 4096 electrodes), and I progressively decreased the electrode density, down to 60 electrodes (inter-electrode distance of 189 μm, electrode density of 19 electrode/mm^2) to obtain a situation comparable with the MEA-60 devices, as previously reported [18]. The obtained results are shown in Fig. 3.10g: the SWI decreases down to a random topology becoming variable and unstable when the number of considered electrodes is less than 100. This last result is referred to the excitatory links and the same analysis was not applied to the inhibitory connections. In fact, such inhibitory links are much less than the excitatory ones, thus leading to an inhibitory topology reconstruction that is strongly influenced by the decimation scheme applied to reduce the number of electrodes.

3.4 FNCCH Results: Discussion and Observations

The computation of the correlation of firing activity in the framework of multiple neural spike trains has been around since the 1960s. For over thirty years, cross-correlation, its generalizations [19], and its homolog in the frequency domain [20] have been the main tools to characterize interactions between neurons organizing into functional groups, or "neuronal assemblies". A common established technique was to build a cross-correlogram (CCH), describing the firing probability of a neuron as a function of time that elapsed after a spike occurred in the other one. Nevertheless, in the literature, there has been no standard definition of the CC and the strength of a connection can be estimated by different means. To make the correlation coefficient independent of modulations in the firing rate, and to provide a basis for the evaluation of the significance in correlation measurements, and in turn, to allow the interpretation within an experiment and comparison between experiments, it is essential to have procedures for correction, normalization and thresholding of the coincidence counts obtained from cross-correlation calculations. Commonly used normalization procedures are related to Normalized Cross-Correlation Histogram (NCCH) [21, 22], event synchronization [23], Normalized Cross-Correlation (NCC—Pearson Coefficient) [24], Coincidence Index of the CCH [2]. Once that a Functional Connectivity Matrix (FCM) is obtained, a thresholding procedure is necessary to discard those values that are not related to putative real connections. All these approaches present advantages and disadvantages but none of them have been applied to reliably identify inhibitory connections on large-scale network from spiking activity. I introduced a filtered and normalized CC based algorithm (i.e., FNCCH) from which thresholded functional connectivity matrices for excitation and inhibition can be robustly obtained.

From the analysis of the data, I identified both small-world and scale free topologies in cortical networks for the excitatory and inhibitory sub-populations. More specifically, I extracted inhibitory subnetworks in cortical and striatal neuronal cultures demonstrating the capability of the method and offering new understanding of neuronal interactions. Finally, the proposed algorithm, while confirming already presented preliminary results in the literature, demonstrates a new way (i.e., through large-scale MEAs and CCH based analysis) to investigate network topology and opens up new perspective for the analysis of multisite electrophysiological recordings.

3.4.1 *Identification of Inhibition*

Generally, by inspecting the CCH we can notice an increase or a decrease of the fluctuations [25]. In some studies it was noticed that the primary effect of inhibition on the cross-correlogram is a trough near the origin, and for this interaction to be visible there must be present a background of postsynaptic spiking against which the inhibitory effect may be exercised (high-tonic firing rate regime) [26, 27]. From some experimental works related to the analysis of connectivity from cortical multi-site and multi-unit recordings [28], reporting the normalized strength of the identified connections it appears that for excitation a good sensitivity is fairly commonly obtained, while the situation is considerably worse for inhibition. This is due to a low sensitivity of CCH for inhibition, especially under conditions of low firing rates [29], as previously demonstrated [26]. The noted difference in sensitivity may amount to an order of magnitude, and it was demonstrated that for inhibition the magnitude of the departure relative to the flat background is equal to the strength of the connection, whereas for excitation it involves an additional gain factor [26]. As a whole, the lack of efficiency in the detection of inhibition simply reflects the disproportionate sensitivity of the analysis tool [30]. In our work, I introduced a cross-correlogram filtering approach (FNCCH) that I developed to overcome the inhibition detectability issue. As Fig. 3.6 shows, the FNCCH is able to detect with very high accuracy the inhibitory links when applied to in silico neural networks with similar dynamics with respect to actual ones. The filtering procedure improves also the detectability of excitatory links resulting in a reshaping of the ROC Curve (Fig. 3.6c) with an increase of both precision (MCC Curve, Fig. 3.6d) and AUC with respect to the standard cross-correlation. However, being a cross-correlation based method, the presented FNCCH has indeed some limitations in the inhibitory links detection that I tried to investigate with the in silico models. The major factor affecting the detectability of inhibition is the variability of the cross-correlogram.

In order to reduce this variability, it is possible to increase the number of coincidences per bin by increasing the bin-width (that is, down-sampling with loss of information in the acquired electrophysiological data), or by increasing the number of events involved (which can be obtained with high firing rate and/or by increasing the recording time) [31]. Another influencing factor depends on the balance of excitatory and inhibitory neuronal inputs (i.e., balanced model) and it is referred to the relative strength between inhibitory and excitatory inputs. In fact, when the neuron is not balanced, excitation is, on average, stronger than inhibition. Conversely, when the neuron is balanced, both excitation and inhibition are strong and thus detection of inhibitory links improves [25, 30, 32]. Starting from the in silico model, I was able to investigate the impact of rates variability on excitation/inhibition detectability, and try to define a reasonable threshold (criterion for detectability [25, 29]). In particular, I varied the firing rate of the inhibitory neurons from 20 to 2 spikes/s while maintaining a firing rate of 2–3 spikes/s for the excitatory neurons. I found that the detectability of functional inhibitory links is preserved with our method down to a firing of about 6 spikes/s and then decreases significantly (data not shown). I also investigated the inhibition identification with respect to the recording time. Starting from 1 h of simulation, I reduced (10 min steps) the recording time and I found that there is a decrease in the inhibition detectability below 30 min of recording (cf. Fig. 3.9f). The obtained results enabled us to apply the FNCCH to in vitro large-scale neural networks, and allowed us to infer topology and functional organization. The described procedure can be also directly applied to Multi Unit Activity (MUA) from in vivo multi-site measurement recordings.

3.4.2 The Emergence of a Scale Free and Small-World Topology

The cortical networks probed with MEA-4k's showed an unambiguous small-world topology. The inhibitory functional links had a SWI equal to 9.2 ± 3.5, higher than the value extracted from the excitatory links (5.1 ± 1.9). Conversely, the cortical networks coupled to the MEA-60 showed a random organization topology (0.21 ± 0.212 for the inhibitory links and 0.38 ± 0.1 for the excitatory ones). The apparent random organizations are due to the low number of recording site of the acquisition system; in fact, it is worth to remember that the SWI is computed by comparing cluster coefficient (CC) and average shortest path length (PL) of the analyzed networks to the corresponding values for surrogate random equivalent networks (same number of nodes and links). From the obtained results, contrarily with what has been recently presented [17], I demonstrated that the emergence of small-worldness, and above all, deviations from a random topology on a reduced set of electrodes (<100 in the case of the MEA-60) cannot be reliably derived or observed in a neuronal population probed by a reduced number of recording sites. Definitively, besides the crucial importance of well-defined statistical tools used for the analysis, it is fundamental to probe network activity by using large-scale microtransducer arrays (i.e., with at least 200 electrodes). As a whole, the issue related to the low number of recording sites should be carefully taken into account when extracting dynamical features as well as organizational principles of complex networks.

Finally, it should be underlined that I limited and focused the work on the CC based methods. I mentioned in the Introduction the widespread and continuously increase use of Information Theory based techniques. Beside the relative novelties of such methods and the good performances (for a review see [33] and references therein), they showed high computational costs and, to our knowledge, the inability to reliably estimate inhibitory connections [2]. Although theoretically, information theory based methods, like Transfer Entropy (TE) and Mutual Information (MI) are, in principle, able to detect inhibitory links, I am not aware of studies consistently reporting a successful identification of inhibitory connections. The problem might be more in the lack of possibility to distinguish between excitatory and inhibitory links, rather than the detection of inhibition, as I will show in the next paragraph.

3.4.3 Comparison with a Transfer Entropy Based Algorithm

In this section, we compare FNCCH with Delayed Transfer Entropy (DTE) [4]. Figure 3.11a shows the ROC curve and the correspondent AUC (Fig. 3.11b) when DTE is applied to an in silico network made up of 1000 neurons (cf., Methods). The ROC curve relative to the total number of links (black curve) displays a shape similar to the one correspondent to the NCCH (black line of Fig. 3.11c, main text). However, if we analyze separately excitatory and inhibitory links, the results change dramatically. If we restrict the detection only to the excitatory links, the DTE ROC curve greatly improves (Fig. 3.11a), together with the AUC value (Fig. 3.11b). The ROC curve related to the inhibitory links (Fig. 3.11a) shows a good TPR value higher than 0.4 in correspondence of a FPR of 0.01. Finally, the ROC curve relative to the

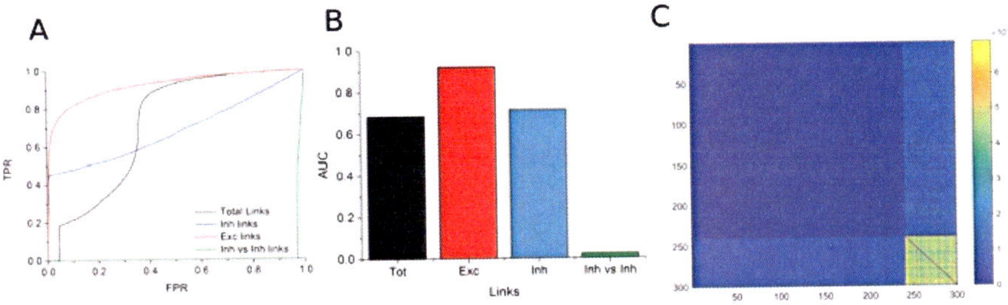

Fig. 3.11 DTE effective connectivity estimation relative to an in silico neuronal network. Effective links are estimated starting from the simulated multi-site electrophysiological activity. **a** ROC curves relative to the total links (black), to the excitatory versus excitatory neurons' links (red), to the inhibitory versus excitatory neurons' links (blue) and to the to the inhibitory versus inhibitory (green). **b** Correspondent AUCs. **c** DTE weighted connectivity matrix. Reproduced from [9]

links between inhibitory neurons (green curve) shows that only false positives are detected, since in the model there are no links among inhibitory neurons. Moreover, these false detected connections correspond to higher values with respect to all the other detected functional links (as confirmed by the connectivity matrix, represented with false color in Fig. 3.11c) with the consequence of worsening the total links detection. Thus, it is indeed possible to "detect" inhibitory links with a TE based method, but is not possible to "identify" (i.e., to distinguish) excitatory and inhibitory links.

3.4.4 *Partial Correlation*

I applied PC to the in silico neural networks described in detail in the appendix. I compared PC to TE and basic CC. In particular, I extracted subsets made up of 60, 120, and 240 neurons. Furthermore, I systematically tested how the aforementioned methods are able to account for different connectivity degrees. To this end, we increased the average connectivity degree of each node of the network model by considering low- (~20%), medium- (~50%) and high-connected (~100%) values.

3.4.5 *Validation of the PC by Means of in Silico Neural Networks*

The performances of each connectivity method were evaluated by means of ROC and ACC curves, averaged over 10 realizations (lasting 10 min) for each network model. The ROC curves of our partialization procedure, observed considering low (L) connected assemblies and low number of nodes (60 MNeu), were very close to the ideal case (i.e. TPR values equal to 1); by increasing the complexity of the neural networks from medium (M) to high (H) connectivity degree, PC maintained higher performances than one-delay TE and CC (Fig. 3.12a, first row). PC showed very good performances also in larger (120 and 240) assemblies at low and medium connectivity degrees (Fig. 3.12a, second and third row), corresponding to reasonable physiological conditions [12]. The extreme situation of high connected networks is considered as limit case to test the performances of the connectivity methods. For this condition, we observed a clear degradation of the partialization approach (Fig. 3.12a, third column). This result might be explained with the so-called "married-nodes" effect in which two generic nodes could be interpreted as direct (i.e., mono-synaptically) connected if they are connected to a third common neuron [34]. These results were confirmed by the AUC analysis that showed PC statistically different (and better) with respect to CC and TE (Wilcoxon rank sum test; Fig. 3.12b) not only in small (60 nodes: $p < 0.001$ at L, M and H connectivity degree) but also in larger networks

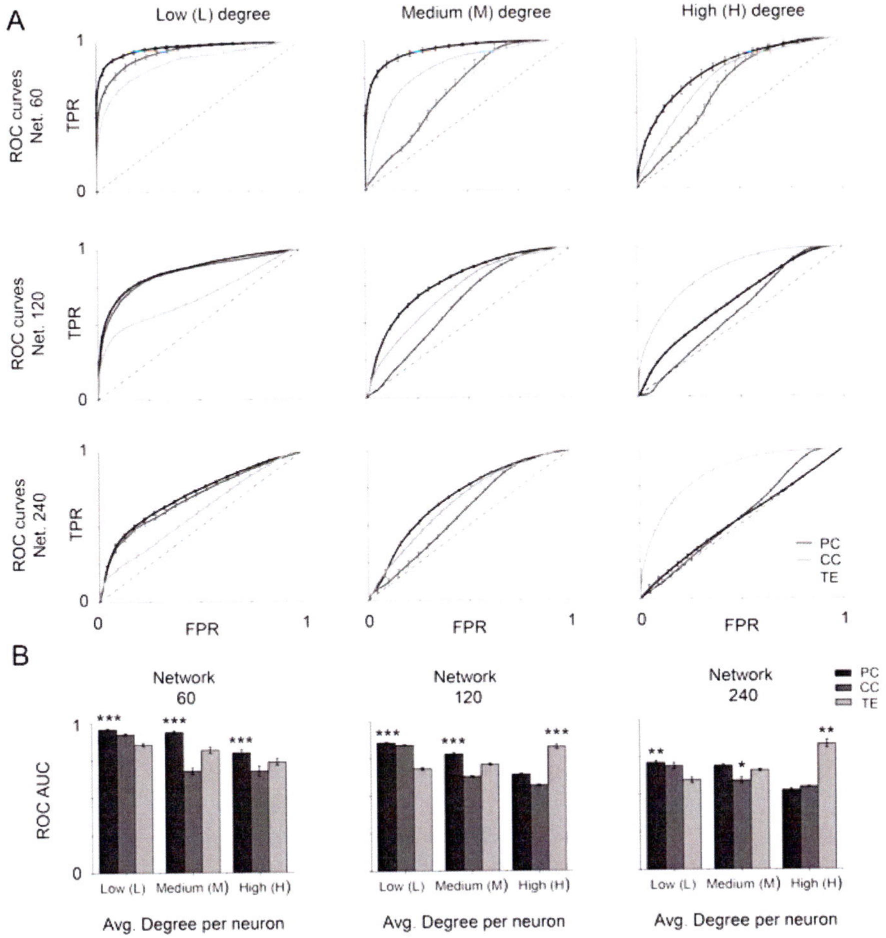

Fig. 3.12 Partial Correlation performances. **a** ROC curves. Mean shape was extracted from 21 simulations; PC algorithm shows significant performances at low and medium connectivity degrees, not only in small (60) but also in larger (120–240) assemblies. Only the high-connected networks, increasing the number of "married nodes", worsen the partialization process. **b** Areas under the curves (AUC). Values of the areas under ROC curves: PC shows very good performances, statistically different respect to CC and one-delay TE (Kruskal–Wallis test. $p < 0.05$, 0.01, and 0.001 identified by one, two and three stars, respectively). Reproduced from [3] with permission from 2016 IOP Publishing Ltd; permission conveyed through Copyright Clearance Center, Inc ©

(120 nodes: $p < 0.001$ at L and M connectivity degree; 240 nodes: $p < 0.01$ and $p < 0.05$ at L and M connectivity degree, respectively).

3.4.6 PC Application to Experimental Data

I also applied the developed partial correlation algorithm on neural networks coupled to the MCS60 low resolution acquisition system.

Modularity Analysis

I first considered data coming from segregated yet structurally and functionally connected neuronal populations, and compared with homogeneous networks (control). The rationale was to further support, with experimental data, whether the partialization approach and the implementations of CC and one-delay TE proposed in this work, were capable of clearly individuating the two segregated populations by means of the modularity index Q [5, 35]. Such an index was extracted by applying first the three methods to n = 8 simulations coming from an interconnected network model, and then to the spontaneous activity of n = 8 cortical assemblies grown in vitro by using a dual-compartment experimental set-up. I quantified whether and to what extent the methods identified the segregation effects. Figure 3.13a, b shows two examples of functional connectivity graphs, obtained by means of PC applied to the two experimental set-ups (i.e., the standard MEA device (Fig. 3.13a) and the dual compartment system (Fig. 3.13b: the black lines indicate the functional connections that cross the physical barrier between the compartments). I analyzed data during the third week in vitro since it is considered a stable and mature stage of development of dissociated cortical networks [36].

The compartmentalization effect was identified by all connectivity methods both in silico (Fig. 3.13c) and in vitro (Fig. 3.13d) models; however the modularity detected by PC was much higher and more significantly different than CC and one-delay TE, both in simulated (Fig. 3.13c; PC modularity ~0.59 $\pm$ 0.01, $p < 0.001$; CC and one-delay TE modularity <0.5 $\pm$ 0.02, $p < 0.01$ and $p < 0.05$ respectively) and experimental conditions (Fig. 3.13d; PC modularity ~0.63 $\pm$ 0.04, $p < 0.001$; CC and one-delay TE modularity <0.5 $\pm$ 0.03, $p < 0.01$ and $p < 0.05$ respectively). This is a clear and additional proof about the actual performances of the connectivity methods used in these experimental conditions.

3.4.7 Connectivity During Development

As application of our partialization method, I characterized the functional-effective connections and I derived the topological structures in developing neuronal cultures (i.e., from the second to the fourth week in vitro). I considered two datasets: the first one composed by n = 7 spontaneous homogeneous cortical assemblies (2 batches), the second consisting of n = 10 chronically stimulated cultures (3 batches) by low-frequency electrical pulses (cf. Sect. 2.1). I estimated functional connectivity by means of the PC method and I evaluated basic graph measures. I normalized them by the average of the same graph indices extracted by 100 randomized surrogates

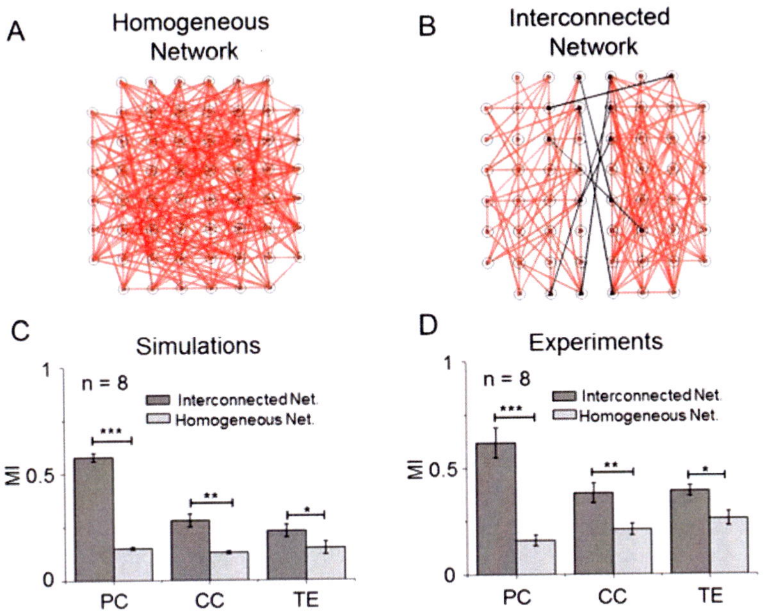

Fig. 3.13 **a** Example of functional connectivity map obtained by applying PC algorithm to a homogeneous cortical network. **b** Connectivity map obtained by applying PC algorithm to a dual compartment network. **c** Modularity Index obtained by applying PC, CC and one-delay TE methods to the homogeneous (grey) and modular cortical network model (dark grey): PC method is a very good choice to detect the compartmentalization effect introduced by the structure of the experimental set-up ($p < 0.001$). **d** Modularity Index obtained by applying PC, CC and one-delay TE methods to the experimental data (homogeneous—grey-, and interconnected network—dark grey-). The results obtained in the computational model were confirmed also during experimental analysis. Reproduced from [3] with permission from © 2016 IOP Publishing Ltd; permission conveyed through Copyright Clearance Center, Inc.

graphs (graph characterized by the same number of links and involved nodes with respect to the experimental data), following the approach devised in [6].

When monitored during development, in vitro cortical cultures started to display relevant spontaneous electrophysiological activity after the first week in vitro. For this reason, I considered the evolution of the cortical networks from the second to the fourth week.

The normalized network statistics (Fig. 3.14a) did not show a significant trend during development (Wilcoxon rank sum test; $p > 0.05$). The increasing entropy values that I found from the second to the fourth weeks in vitro, instead, provided an important contribution to identify a significant emergence of a random structure (Fig. 3.14c, red line; Wilcoxon rank sum test; $p < 0.001$), supporting the hypothesis that cortical assemblies had a tendency to integrate rather than to segregate.

In addition to spontaneous-evolving networks, I studied how the chronic delivery of electrical stimulation could affect the development of functional links. I considered the development of n = 10 cortical networks during the pre-stimulus phases, from the

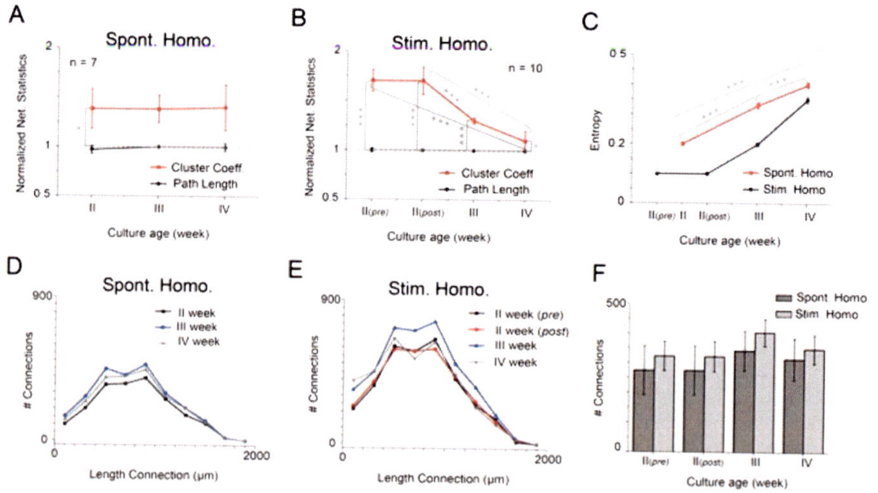

Fig. 3.14 **a** Cluster Coefficient (red line) and Path Length (black line) analysis of 7 spontaneous homogeneous networks during development. **b** Cluster Coefficient (red line) and Path Length (black line) analysis of 10 stimulated homogeneous networks during development. **c** Entropy of spontaneous (red) and stimulated (black) homogeneous networks. **d**, **e** Link distributions during development. **f** Histogram of number of connections extracted from spontaneous (dark grey) and stimulated (grey) homogeneous networks. Reproduced from [3] with permission from © 2016 IOP Publishing Ltd; permission conveyed through Copyright Clearance Center, Inc

second to the fourth week by studying how the topological structures changed from the pre-stimulus to post-stimulus phase during the second week. By evaluating the cluster coefficient, I observed that young cultures (II week in vitro) exhibited a more segregated structure. In the following weeks (IIIrd and IVth) the cultures exhibited the emergence of an integrated structure. This is clearly visible in Fig. 3.14b with a decrease of the clusterization effect while keeping constant the normalized values of Path Length. I observed a significant integration trend during development (Wilcoxon rank sum test: II week pre or post stimulus/III-week pre-stimulus $p \ll 0.01$; II week pre or post stimulus/IV week pre-stimulus $p \ll 0.001$). Moreover, the cluster coefficients that I found were significantly different from those extracted by analyzing random network models (Wilcoxon rank sum test: II week pre or post stimulus/random case $p \ll 0.001$; III-week pre-stimulus/random case $p \ll 0.001$; IV week pre-stimulus/random case $p < 0.05$). These results were confirmed also by the increasing Entropy levels, evaluated for each network node having non zeros cluster coefficient values (Fig. 3.14c, black line). Our analysis shows how the stimulation promoted a significant change in the network topology favouring the emergence of a random functional structure as indicated by the decrease of the normalized Cluster Coefficient values while keeping the Path Length measures on the maximum integration levels (levels equal to 1) during development (Fig. 3.14b). Moreover, I observed that the stimulation did not have any effect on the connections length, since their distributions did not change from the spontaneous to the stimulated homogeneous networks (Fig. 3.14d, e). I did not find significant differences at level of number of links (Fig. 3.14f). Finally, I also characterized the functional topological structures of the same spontaneously developing and electrically stimulated networks by means of CC and one-delay TE methods. The obtained results did not show significant trends during development [3]. Altogether these results support the idea that PC can be conveniently applied to estimate structural connections from functional-effective links and that is capable to discriminate subtle changes in the network topology.

3.5 Algorithm Optimization

The functional connectivity analysis is a quite demanding computational process, and all of the previous described connectivity methods for analyzing multiple spike trains rely on quantities that need to be computed through intensive calculations [4]. In order to reduce the time requested for the computation and to improve the Random Access Memory (RAM) usage, I decided to combine informatics tool for the implementation of the statistic methods, with a computational logic approach based on a redefinition of the connectivity algorithms to be tailored for spike train data. In this section, I will describe the latter approach, while in the next one, I will introduce and define the informatics instruments as well as the informatics tools that I developed for the complete analysis of dynamics and functional connectivity of spike trains data. The computational logic approach modifications allowed a very drastic reduction of the time requested for the computation and the RAM usage. In fact, I implemented a C# optimized version of the transfer entropy algorithm using informatics optmization strategies to reduce the time needed for the computation. I used an in silico neural networks with 500 Izhikevich-based neurons simulated for 1 h to test the developed algorithm. That implementation required 42 h to provide the full 500 $\times$ 500 connectivity matrix with a 4 cores i7 2.5 GhZ, 16 GB of RAM and solid state disk. Such a computational time was not optimal in the optic of analysing neural networks coupled to MEA-4k with 4096 electrodes. Spike trains are very sparse data and, as a consequence, the possibility to consider only the non-zeros value in the algorithm computation, can drastically reduce the requested time for computation. The chosen strategy is based on the efficient use of time stamps (i.e., the index of a sample correspondent to a spike) as input and working data. In fact, in the in silico networks I used for the testing, the firing rate of neurons are around 2–3 spikes/s, meaning that there are totally 7200–10,800 spikes with respect to 3600,000 total number of samples (sampling frequency set at 10 kHz), with a reduction of the data dimensionality of more than 3 orders of magnitude. In the next paragraphs, I will describe the logic optimization and implementation of FNCCH, TE and JE algorithms.

3.5.1 FNCCH: Algorithm Optimization

The optimization procedure for the FNCCH is based on time stamps relative to the occurrence of a spike on a specific electrode. The block diagram and pseudocode depicted in Fig. 3.15 shows the computational operations executed by the optimized CC algorithm (FNCCH). We start from the first bin containing a spike in the target train. The binning procedure is directly performed on the time stamps. For each pair of neurons, starting from the first spike of the target train, we slide the time stamps of the reference electrode to find the first spike whose correlation window contains the target investigated spike. Then, we continue to move over the target train to build the entire cross-correlogram (for that reference spike). When the correlation window for the reference spike is completed (i.e., when we have counted the number of spikes for all the bin of the target spike train), we move to the next spike of the reference train, and re-iterate the procedure starting from the first target spike into the correlation window, centered in the current reference spike. Thus, we normalize the CC and repeat all the aforementioned operations for the other electrodes. An important optimization strategy exploits the symmetry of the CC function allowing considering only half of the electrodes for the computation. Moreover, we choose, as target train, the one presenting the smallest number of spikes to reduce the number of operations for each pair. Once the NCCH is obtained, we apply the filtering operation described by Eq. (2.2) to compute the FNCCH values. Finally, we take the maximum absolute value as estimation of the correlation between the two electrodes. If it is

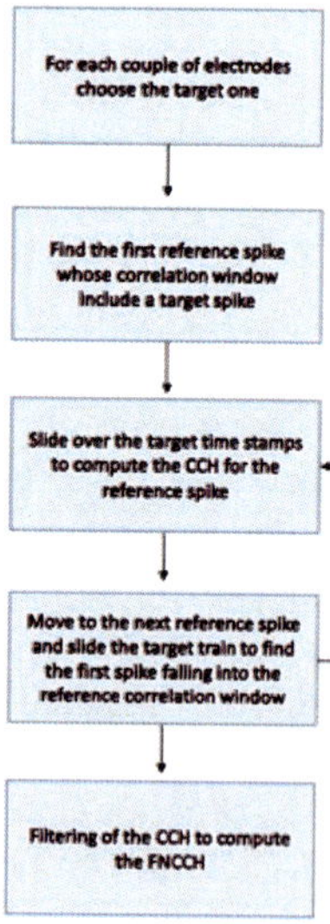


```
for electrode_i in electrodes
    for electrode_j in electrodes
        if length(timestamps_electrode_i > length(timestamps_electrode_j)
            target = timestamps_electrode_i;
        else
            target = timestamps_electrode_j;
        end if
        for spike_target in target
            spike_reference = find(spike_target);
            while corr_window_ref_contains(target)
                bin_of_target = spike_reference - spike_target +
                corr_window(spike_reference)
            end while
            spike_reference = next(spike_reference)
            spike_target = first_spike_target_in_corr_window(spike_reference)
        end for
    end for
end for
```

Fig. 3.15 Schematic representation and description of the algorithm to obtain the FNCCH. Reproduced from [9]

negative, the found connection is likely to be an inhibitory link, otherwise the found connection is likely to be an excitatory link.

3.5.2 *Transfer Entropy Algorithm Optimization*

Figure 3.16 schematically represents and describes the main operations of the implemented TE algorithm. First of all, the customized algorithm performs a binning procedure (with a bin width set by the user via the GUI) directly on the time stamps. Then, we start the computation of the information transfer among each couple of electrodes. We compute the TE order, that is equal to *x_order* (i.e., number of bins in the reference train) + *y_order* (i.e., number of bins in the target train) + 1 (past bin in the reference train). At this stage, we build a pattern vector with TE order elements initialized with *x_order* bins of the reference train, the one past bin of the reference train, and the *y_order* past bins of the target train. The number of all the possible patterns is 2^{TEorder}, and we use a decimal code (0, 1, 2,..., TE order) for coding each binary pattern. Thus, starting from the first element of the pattern vector, we select the minimum bin (i.e., minimum spike time stamp value) and initialize the code to the simplest pattern (1 in that bin, and 0 for all the others). Then, we substitute this element of the pattern vector with the next spike (from the correspondent train) and

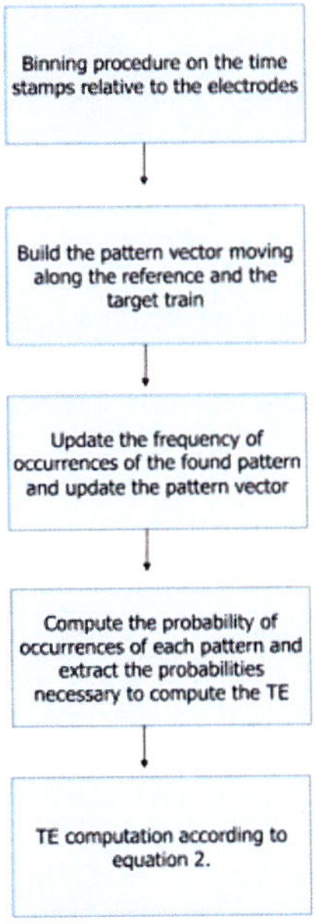


```
For delay in delays
    For electrode_i in electrodes
        For electrode_j in electrodes
            Pattern = build_pattern_i_j;
            While (minimum_bin < end_time)
                minimum bin = minimum(Pattern);
                code = 1;
                Pattern(minimum bin) = next_spike(electrode_containing_min_bin);
                While (previous_pattern.Contains(minimum bin))
                    update(code);
                End While
                frequency_of_code(code)++;
            End While
            frequency_of_code(Null_Code) = end_time - sum(frequency_of_code);
            Probability_of_occurrence = frequency_of_code/number_of_occurrences;
            TE_i_j (delay) = compute_transfer_entropy (Probability_of_occurrence);
        End for
    End for
    TE_i_j = max_over_delays(TE_i_j)
End for
```

Fig. 3.16 Schematic representation and description of the algorithm to obtain the TE. Reproduced from [4]

select again the minimum bin indifferently from the reference or the target train. If this bin belongs to a common pattern with the precedent bins the code is updated and increased, otherwise, a vector of length TE order, representing the frequency of occurrence of each code is updated and the code is re-initialized to the simplest pattern for the updated minimum bin. We repeat all these operations until the number of spikes of the reference or the target train is ended. The null code (related to the pattern formed by only empty bins) is obtained as the total number of recorded bins minus the sum of all other patterns frequency of occurrences. At this stage, we use the frequency of occurrences to extract each probability of occurrences and we compute the TE value among the analyzed electrode by applying Eq. (2.12). For delayed transfer entropy, I used a multithreading implementation to have a different thread for every different delay. At the end of the computation, I save the maximum value of TE and the correspondent delay.

3.5.3 *Joint Entropy Algorithm Optimization*

Figure 3.17 schematically describes and represents in form of pseudocode the main operations executed by the customized JE algorithm's implementation. Considering the first spike of the reference train, we slide the target train to find the first spike

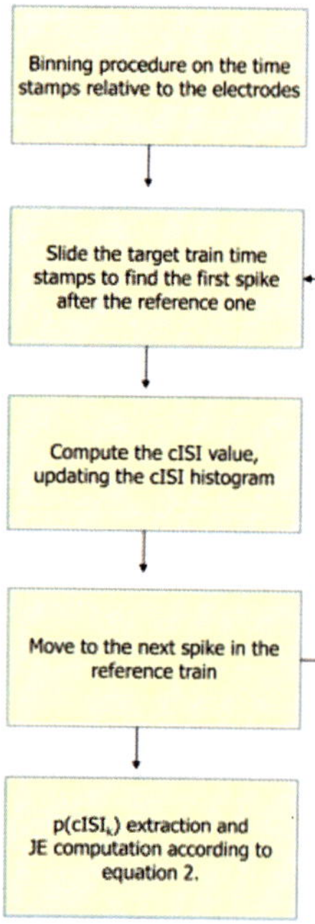


```
For electrode_i in electrodes
    For electrode_j in electrodes
        For spike_x in timestamps_electrode_i
            spike_y = find_first_spike_after (spike_x)
            cISI = spike_y - spike_x;
            cISI_histogram(cISI) ++;
        End for
        Probability_of_cISI = cISI_histogram/total_number_of_cISI;
        JE_i_j (i,j) = compute_joint_entropy (Probability_of_cISI);
        Delay(i,j) = most_frequent_cISI_width();
    End for
End for
```

Fig. 3.17 Schematic representation and description of the algorithm to obtain the JE next to the reference one. We compute the cISI value and update the cISI histogram (i.e., the counting of cISI of specified sizes). We iterate such a procedure until all the reference's spikes have been analyzed. Finally, we compute the cISI probability normalizing the number of cISI of a specific width for the total number of cISIs. The application of (15) allows the computation of the JE between the analyzed electrodes.

3.6 Software Development for Functional Connectivity Analysis

Recent advances in multichannel extracellular recording techniques have made possible the simultaneous recording of the electrophysiological activity of thousands of neurons, with in vitro arrays that can contain up to thousands of microelectrodes [37–41]. Such relevant advancements in the technology have made routine the acquisition of massive amounts of data. Moreover, quite often the experimental protocols researchers perform require long recordings (e.g., tens of minutes, hours) which generate a huge amount of raw data. For the above reasons, new computational strategies and informatics toolboxes are requested for optimizing the management and analysis of the acquired data. When I started my PhD, to the best of my knowledge, there was no available dedicated software that put together a set of different functional connectivity analysis methods, with a friendly Graphical User Interface (GUI) and appositely tailored to be applied on spike train data. For this reason, I decided to develop a user-friendly toolbox [42] in order to provide the researchers community with a powerful tool to perform functional connectivity analysis on in-vitro neuronal networks coupled to standard and high-density MEAs, while guaranteeing computational efficiency and high accuracy.

3.6.1 Main Graphical Interface

I designed and developed ToolConnect taking care of the user-friendliness. According to this, our software package has a Graphical User Interface (GUI), which permits also to inexperienced users to perform functional connectivity analysis, to graphically represent the results, without knowing the details of the algorithms and the organization of the software's code. Figure 3.18 shows in form of block diagrams the organization of ToolConnect, while Fig. 3.19 shows an example of the graphical tools provided to the users by ToolConnect, and inherited by SpiCoDyn.

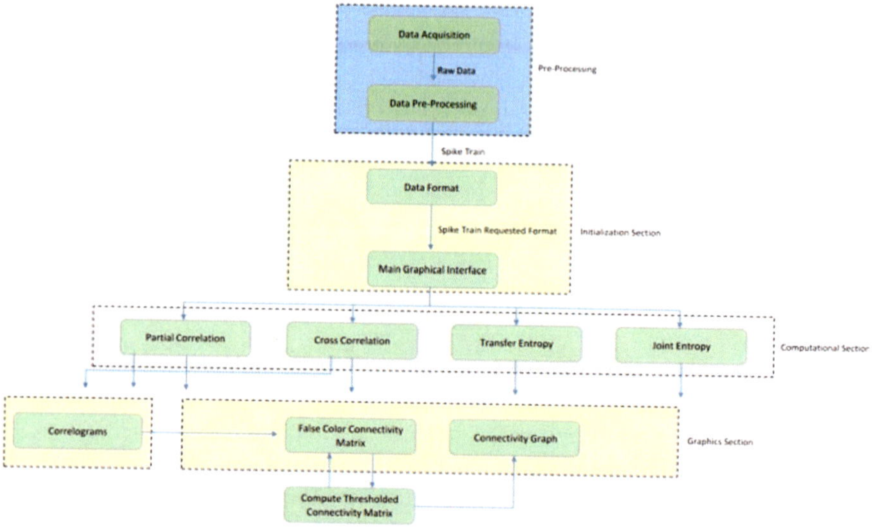

Fig. 3.18 Schematic overview of ToolConnect. The Functional blocks show the computational and graphical tools embedded in the software package. The flow chart starts with the pre-processing section; it includes the data acquisition procedure and all the other operations necessary to create the spike trains, which are the software's input data. Reproduced from [42]

3.6.2 *Example of ToolConnect's Application to Experimental Data*

3.6.2.1 Spontaneous versus Stimulated Networks

I applied the cross-correlation algorithm (in the frequency domain) to cultured hippocampal neural networks (see Supplementary Material for cell culture) during both spontaneous and stimulus-evoked activity. Electrical stimulation was performed by applying bi-phasic voltage stimuli delivered at the frequency of 0.2 Hz, with a peak-to-peak amplitude of 2.0 V, from a single electrode. Analysis was performed on recordings lasting 10 min (sampling frequency of 10 kHz). Figure 3.20 shows the connectivity graphs and the connectivity matrices obtained for the spontaneous conditions (Fig. 3.20a, d) and stimulation phases relative to a stimulation phase from channel 45 (Fig. 3.20b, e) and 21 (Fig. 3.20c, f). All these graphs and matrices were obtained by thresholding the CM using a threshold equal to $\mu + 2 \cdot \sigma$, where μ and σ are the mean value and the standard deviation of the CM's elements, respectively (cf. *hard threshold*). After the thresholding procedure, during spontaneous activity, I detected more links (about 90%) than during stimulated activity (Fig. 3.21c). These functional links involve a larger number of electrodes (about 53%) in the spontaneous condition with respect to the stimulated one (Fig. 3.21d). Moreover, the correlation values obtained for the spontaneous activity are lower but more homogeneous than the values corresponding to the stimulated one (maximum values' difference of

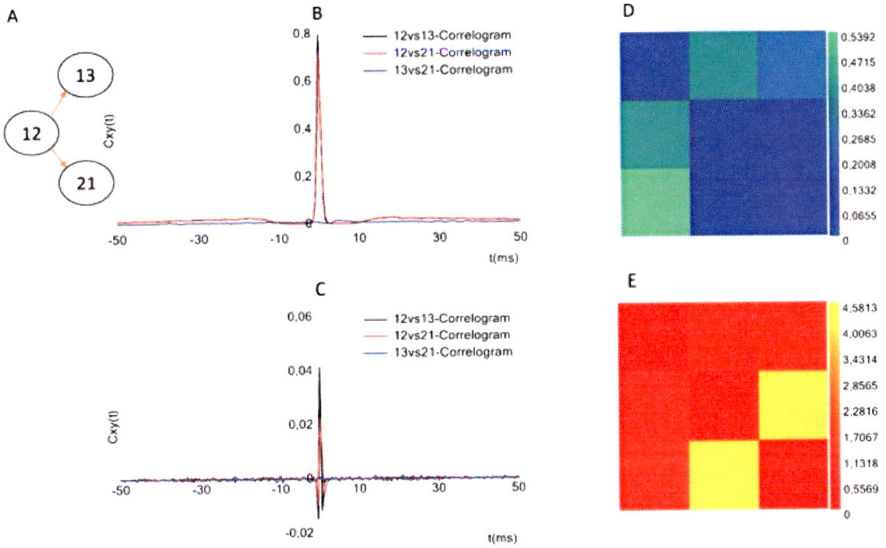

Fig. 3.19 Example of ToolConnect's graphical output. **a** Sketch of connections among three neurons in a simulated neuronal network. **b**, **c** Cross-correlograms computed by using the frequency and the time approach respectively. **d** Partial correlograms computed among the three electrodes. **e**, **f** Correspondent TE and the JE matrices. The peak appearing for the cross- and partial correlograms 12vs13 and 12vs21 corresponds to a connection found between the aforementioned electrodes. In the same way, the TE presents the highest values for the couples (12, 13) and (12, 21) and the joint entropy shows the smallest values for the same couples; meaning that these methods found a connection between the aforementioned electrodes. Regarding the electrodes 13 and 21, instead, the absence of peak in the cross-and partial correlograms, the low values for TE and the high values for JE suggest the absence of a connection, as the sketch of the panel **a** shows. Reproduced from [42]

0.16, standard deviation 0.008 versus 0.045 for spontaneous and stimulated activity respectively, Fig. 3.20d, e, f). I also evaluated the in- and the out-degree distribution. We can observe that during spontaneous activity, the in- and the out-degree for the different electrode are almost equally distributed among the active electrodes (Fig. 3.20g). During stimulus-evoked activity, the stimulated electrode showed an increased number of outgoing connections, reaching a difference of 14 links with the other electrodes (Fig. 3.20h, i). In the case corresponding to the stimulation of electrode 21, we had only one electrode with more than 3 outgoing connections. When stimulating the electrode 45, I found only one electrode with more than 4 outgoing-ingoing connections. In the spontaneous condition, instead, there were 35 and 28 electrodes with an out-degree of connections greater than 3 and 4 links, respectively. Finally, I used two metrics form graph theory (namely, cluster coefficient and path length) to evaluate the topological characteristics of the neural networks, in the two different experimental conditions. Figure 3.21a, b show the results relative to the cluster coefficient and path length, respectively. I could observe that the path length is not affected by the electrical stimulation (i.e., same values for the spontaneous and the

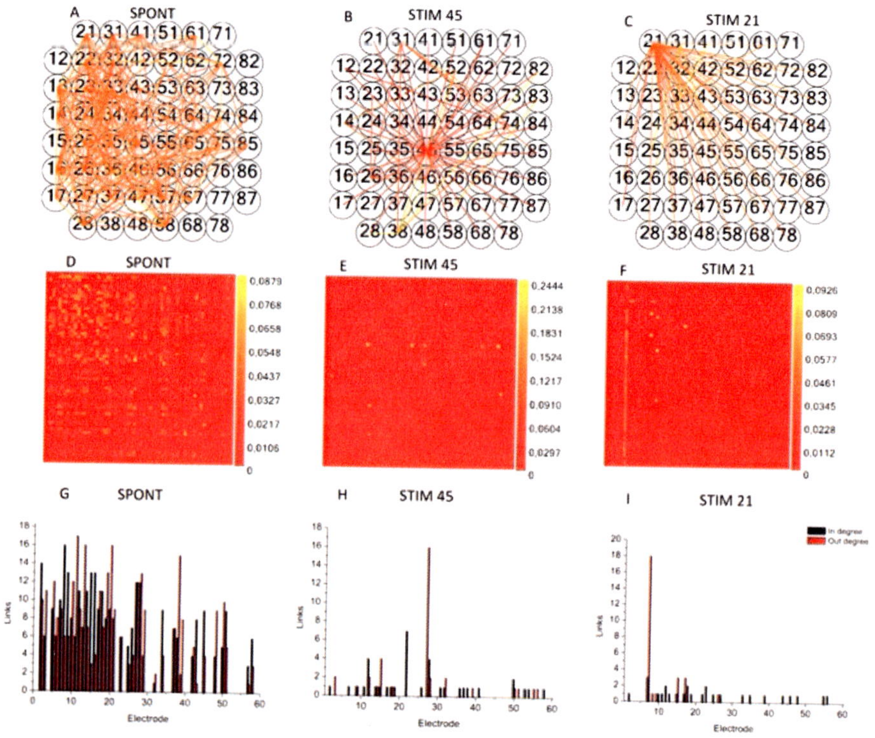

Fig. 3.20 ToolConnect's application to mature hippocampal assemblies. ToolConnect's application to mature hippocampal assemblies. Cross-correlation algorithm was applied to hippocampal networks in spontaneous and stimulus-evoked conditions. **a**, **d**, **g** connectivity graph, connectivity matrix and degree distribution for spontaneous activity. **b**, **e**, **h** connectivity graph, connectivity matrix and degree distribution for stimulated activity (site of stimulation, electrode 45). **c**, **f**, **i** Connectivity graph, connectivity matrix and degree distribution for stimulated activity (site of stimulation, electrode 21). Reproduced from [42]

stimulus-evoked activity). On the contrary, I found a higher cluster coefficient for the spontaneous conditions than for the stimulated ones. This is probably due to the effect of the stimulation on the global network's dynamics: the stimulation empowers the outgoing connections from the stimulated electrode, thus, these connections become stronger, and correspond to higher cross-correlation values. We can observe that, the connectivity graph for the stimulation of the electrode 21 and 45 are quite different. In particular, the outgoing connections of the electrode 45, are weaker than the ones relative to the electrode 21 (Fig. 3.20b, c). Moreover, the degree distribution relative to the electrode 45 shows a higher variability in the number of in-coming and outgoing connections than the one relative to the electrode 21. Figure 3.21e, f shows the population Post Stimulus Time Histogram (PSTH) relative to the two different sites of stimulation (i.e., the PSTH averaged over all the electrodes); we can observe that the number of evoked spikes by the stimulation of channel 45 (Fig. 3.20e) is lower than the number correspondent to the stimulation of the electrode 21 (Fig. 3.21f).

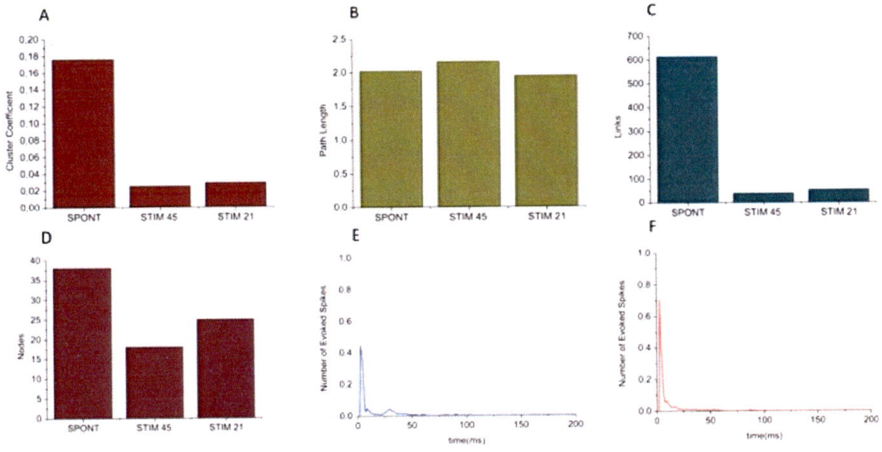

Fig. 3.21 Graph Theory's analysis of hippocampal neural assemblies. **a** Cluster coefficient. **b** Path length. **c**, **d** Number of links and number of neurons found after the thresholding procedure, respectively. **e** PSTH relative to the stimulation of electrode 45. **f** PSTH relative to the stimulation of electrode 21. Reproduced from [42]

It is possible to ascribe that the electrode 45 is less involved in the dynamics of the analyzed network with respect to the electrode 21. Reproduced from [42].

3.6.2.2 SpiCoDyn

I implemented an automated and efficient open-source software for the analysis of multi-site neuronal spike signals. The software package, named SpiCoDyn, has been developed as a standalone windows GUI application, using C# programming language with Microsoft Visual Studio based on.NET framework 4.5 development environment. Accepted input data formats are HDF5, level 5 MAT and text files, containing recorded or generated time series spike signals data. SpiCoDyn processes such electrophysiological signals focusing on: spiking and bursting dynamics and functional-effective connectivity analysis. In particular, for inferring network connectivity, a new implementation of the transfer entropy (TE) method is presented dealing with multiple time delays (temporal extension) and with multiple binary patterns (high order extension). SpiCoDyn is specifically tailored to process data coming from different MEAs setups, guarantying, in those specific cases, automated processing. The optimized implementation of the Delayed Transfer Entropy (DTE) and the High-Order Transfer Entropy (HOTE) algorithms, allows performing accurate and rapid analysis on multiple spike trains from thousands of electrodes. The new developed software SpiCoDyn inherits the overall architecture from ToolConnect[42], no more supported, and includes new relevant features such as spike detection, first-order statistics for spikes and bursts, fully revisited TE based algorithms, and topological analysis. Specifically, it includes a brand new temporal

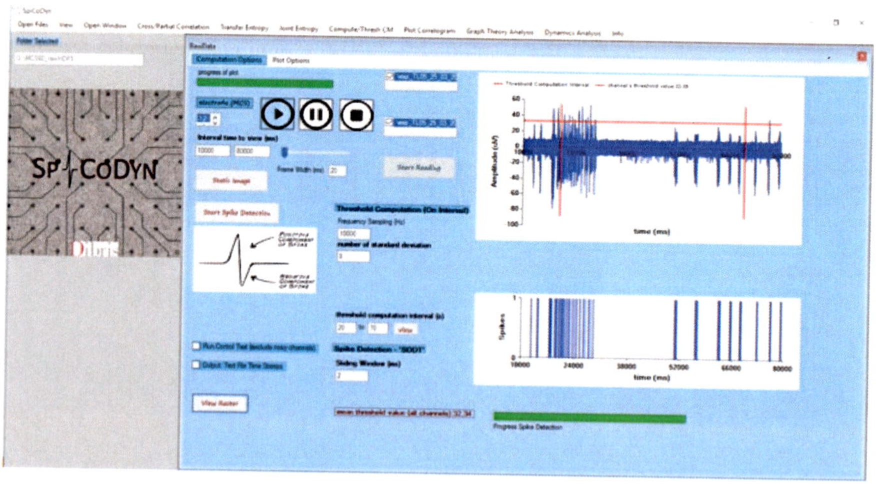

Fig. 3.22 Screenshot of the main interface of SpiCoDyn. Reproduced from [4]

extension of the TE dealing with multiple time delays and orders higher than one [2, 43]. Figure 3.22 displays a screenshot of the main raw data processing user interface providing all the functionalities that will be described in the next sections.

3.6.3 Multi-threading Implementation

SpiCoDyn is a Multiple Document Interface (MDI) windows form application. A father form is a frame in which all the embedded computational and graphical tools' interfaces are opened and displayed. I implemented each of these interfaces independently from the others, as a windows form. A friendly GUI and a complete set of feedback information are fundamental in the definition of the toolbox. According to this, each windows form has a multi-thread implementation: one thread of the application is used to update the graphical interface and another one executes the effective code of the selected connectivity method. In this way, it is possible to update a progress bar indicating the state of the application and other controls reporting useful information (e.g. the start time of the algorithm and the percentage of progress), and to make this interface accessible any time in order to resize, minimize or move the relative window. Some of the threads were directly implemented at code level, while other were implemented by use a.Net Framework 4.5's powerful embedded control to handle the multi-threading application: the backgroundworker. This object provides the user with the methods *DoWork*, *RunWorkerCompleted* and *ProgressChanged*. The first one contains all the code that must be executed in the thread (in our case the connectivity method's code); the second one indicates the operations to run after the conclusion of the thread. The last one indicates the operations to do when a change in the progress occurs (e.g., the update of a progress bar). When performing functional connectivity analysis, the need of a large amount of RAM and the high computational time of execution are the two main issues to manage. The aforementioned multi-threading implementation permits the various connectivity methods to be performed simultaneously on the different threads of the different available CPUs, significantly reducing the computational requested time.

3.6.4 Input Data Processing and Conversion Procedures

Researchers need a reliable format for exchanging large datasets for specific data analysis. A well-recognized possibility of a common file format is based on the HDF5 (Hierarchical Data Format) structure [44]. HDF5 format (structured binary format) allows fast loading and low processing times when dealing with huge amounts of data [45].

For this reason, I adopted this standardized raw input data format, making our tool compatible to deal with data coming from different acquisition systems. In fact, most of the acquisitions system's related software provides tools and packages to convert the specific format or to save the acquired electrode's raw data into a standardized level-2 HDF5 format (e.g., "DataManager" from MCS Systems—"mcd" format, "BrainWaveX" form 3Brain—"brw" format). However, SpiCoDyn offer the possibility to read also the level 5 MAT-files format (v5-v6-v7, The Mathworks, Natick, US) and convert them into textual or binary files, which are the supported formats for spike trains files to be used as input for the spike analysis and connectivity methods. Such a conversion procedure is done by using dedicated open-source.NET libraries and adopting an optimized implementation strategy to extract the structured data elements values. More specifically, I implemented a procedure to iteratively move through the file by finding a tag and then reading ahead in memory the specific number of bytes until the next tag. The conversion procedure offers also the possibility to join spike trains related to distinct phases of the same experiment, provided in separate level 5 MAT-files. Figure 3.23 graphically illustrates the MAT-files format providing details about the structured elements. Finally, SpiCoDyn guarantees the compatibility with spike trains format data generated by SpyCode [46]: a previously developed software from our research group. The main raw data pre-processing features offered by SpiCoDyn includes: (i) electrode's raw signal visualization with a time-varying plot of the selected electrodes in a specific interval; (ii) selectable algorithms of spike detection (Cf., Sect. 2.3.1); (iii) raster plot of the identified spiking activity while the electrode's raw signal is being displayed. Figure 3.24 represents a block diagram of SpiCoDyn's functionalities and process flow, schematically describing the operations performed by the user to select the input data files, to choose the analysis method to perform, and to set the input parameters. It also depicts the operations performed by SpiCoDyn's modules, related to the input raw data management, conversion, spike detection and spike trains analysis (dynamics and spiking/bursting

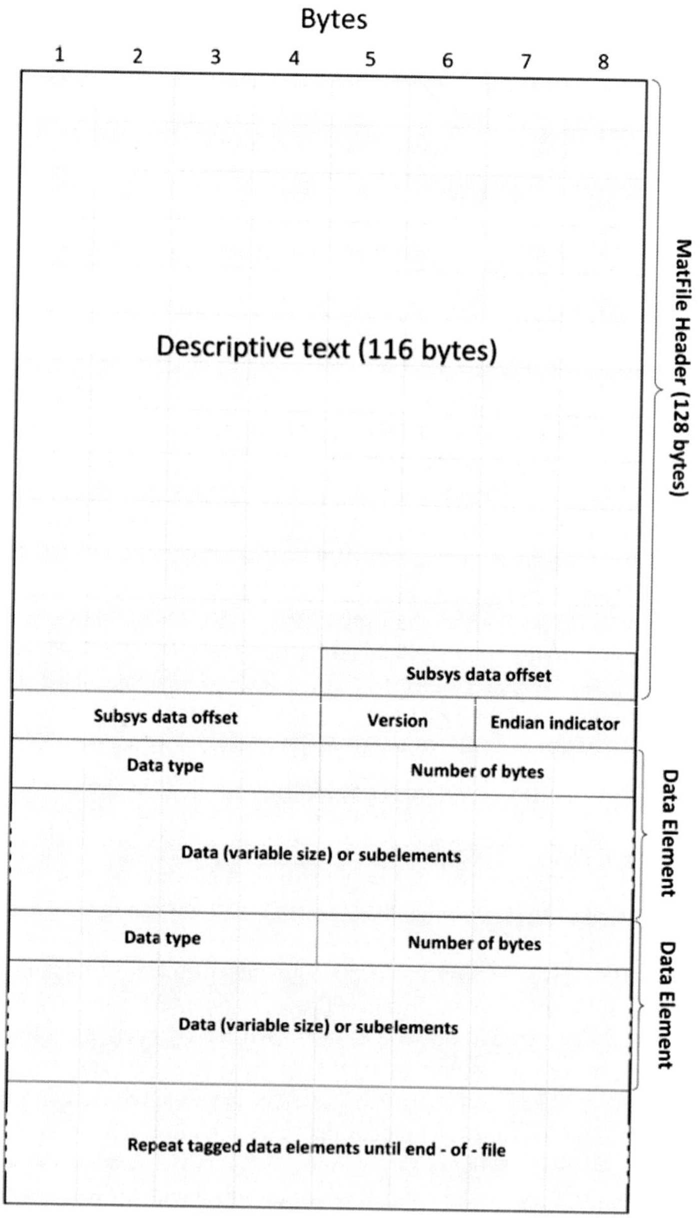

Fig. 3.23 Matlab Level 5 MAT file-format. Reproduced from [4].

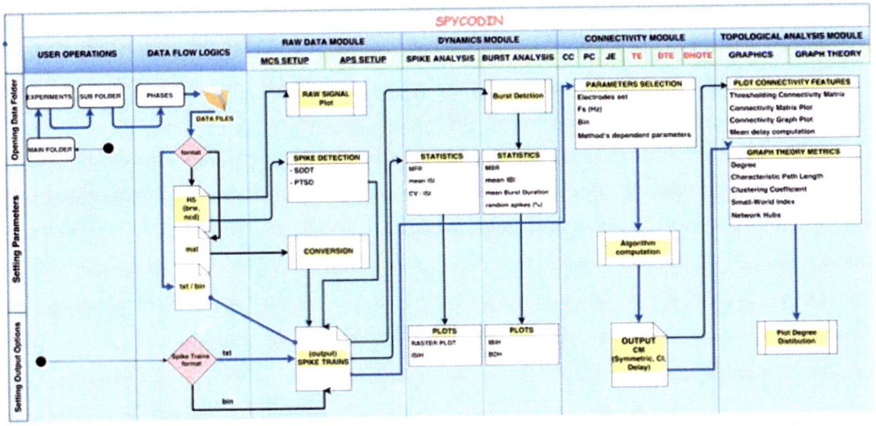

Fig. 3.24 Functionalities block diagram and process flow chart describing the flux of operations performable using SPICODYN. The blue line identifies a possible path to perform a complete analysis starting from the raw data in HDF5 format, and performing spike detection, spiking and bursting analysis as well as connectivity analysis. Reproduced from [4]

statistics, functional-effective connectivity analysis, connectivity matrix and graphs visualization, topological features analysis and graph theory metrics).

3.6.5 *Spiking and Bursting Features Analysis*

The creation of huge amounts of data from simultaneous recording of many (thousands of) neurons with HD-MEA acquisition setups implies the development of efficient methods and specific tailored algorithms to analyze such neuronal spike trains. Spike detection methods, applied to recorded electrodes raw signals, rely on quantities that require intensive calculations. SpiCoDyn's interface provides the user with commonly used algorithms to perform spike detection, and some subsequent statistical analysis on the extracted spike trains data.

3.6.6 *Spike Detection*

The current version of SpiCoDyn implements two spike-detection algorithms: "Spike Detection Differential Threshold" (SDDT) and "Precision Timing Spike Detection" (PTSD).

I developed a dedicated form that provides the necessary fields to specify all the required input parameters. The spike detection algorithms involve a common step related to the threshold computation. I implemented two different strategies to compute distinct threshold values for all the channels:

(i) "automatic" sliding-window threshold computation procedure: the user sets the number of positions of the sliding window (k: typical values between 10 and 20), the percentage of coverage of the total signal length (or the complementary sliding window length—w), and the multiplication factor (n) for the computed standard deviation (SD) of the analyzed signal's portion. The SD value is evaluated for all the windows positions (k), and the minimum value represents the basal noise threshold value. The threshold on every channel can be written as:

$$threshold_{channel\_i} = n * min_k\{SD(channel\_i\_samples(w_k))\} \quad (3.1)$$

(ii) "manual" user-set interval threshold computation procedure: the user sets an interval for the threshold computation on the basis of the signal's amplitude trend by visual inspection. The threshold value for every channel corresponds to the SD value of that precise portion of the signal, multiplied by the SD factor n. Taking into account the setup-dependent signal's amplitude and variability, the automatic procedure for the threshold computation, when applied to 3Brain data, uses the mean SD value computed in all the sliding window positions instead of the minimum SD value. In both methods, I implemented a preliminary control (of the noise levels) procedure, to exclude from the spike detection those channels that exhibits higher levels of noise with respect to a maximum admitted value. In the case of the 3Brain data, this procedure is based on a maximum threshold value that is set by the user. In the case of the MCS data instead, the control is performed only with the "manual" threshold computation option, and the maximum threshold value is computed on a "reference" electrode (set by the user).

I implemented a version of the SDDT procedure following the algorithm as described in [47]. The PTSD procedure is implemented on the basis of the specifics described in [48]. Both procedures have been adapted and customized to deal with multiple electrodes raw data in the HDF5 file format. I took specific care to guarantee an optimized memory data management and to reduce the computational costs. To achieve this, in the worst case of data coming from the 4096 electrodes from the 3Brain setup (sampling frequency of 9043 Hz), I implemented a memory buffer that stores only a short portion of 10 s (i.e., 10 $\times$ 9043 samples per buffer) of the data file in an iterative manner. I also applied a multi-threading approach to parallelize spike detection for different electrodes on different threads.

3.6.7 Burst Detection

A burst consists of a sequence of spikes separated by short time intervals within the sequence, and by a relatively long interval between two spikes belonging to two different sequences [49]. However, the definitions of bursts and burst detection methods differ among studies. Commonly accepted analysis tools employ burst detection algorithms based on predefined criteria [50]. The specific parameters of the methods adopted to identify bursts from MEA recorded spike trains are usually tailored according to the experimental conditions, and generally differs in the inter spike interval (ISI) thresholds definition and usage and in the minimum number of consecutive spikes for burst identification. Some studies use the average ISIs of the measurements [51], average firing rates or logarithmic histogram of ISIs [52] to calculate an ISI threshold for detecting bursts. In the current version of SpiCoDyn, I implemented the detection of bursts as sequences of at least N consecutive spikes spaced less than a chosen ISI threshold value, according to [53]. The two threshold parameters are provided directly by the user through the interactive burst detection interface. The first one is the maximum ISI for spikes within a burst. The second one is the minimum number of consecutive spikes belonging to a burst (N). Our algorithm applies an efficient strategy to scan the spike train time stamps values and calculate the ISIs while at same time evaluating the threshold conditions. Thus avoiding the need to scan the entire sparse binary spike train improving the computational performances and data storage resources usage.

3.6.8 Spiking and Bursting Statistics

Different metrics, such as overall spiking activity, firing rates, bursting frequency and duration, can be used to have a first overview of the dynamic behavior of a neuronal network. Indeed, such simple metrics are extensively applied when MEAs recorded data are used for neuro-toxicology and neuro-pharmacological studies [54, 55]. In SpiCoDyn, I implemented several spiking and bursting statistics to characterize the dynamical features exhibited by the underlying neuronal networks.

(i) Spiking activity (Raster Plot) and spiking statistics: SpiCoDyn provides the computation of the single electrode's firing rates (FR) and the overall network's mean firing rate (MFR), as well as the evaluation of the number of "active" electrodes (i.e., with FR > FR_threshold_value). After the spiking statistics computation, the user can choose the requested time intervals to plot the raster plot trough the dedicated interface.

(ii) Inter Spike Interval (ISI): is defined as the time interval between consecutive spikes and it is a commonly used metric in the analysis of neuronal recordings. SpiCoDyn provides the network's mean and relative standard deviation of ISI, and the ISI coefficient of variation (CV). It also provides the ISI histogram (ISIH) computation and plots for the whole set of electrodes. The latter is constructed evaluating the time (samples) intervals between every couple of consecutive spikes while binning at a user specified bin width:

$$ISIH(bin_k) = \sum_i \left[k - 1 \leq \frac{(t_i - t_{i-1})}{w} < k\right], k = 1, \ldots, m; \; i - (2, N) \quad (3.2)$$

where w is the bin width, k is the k-th bin index, m is the maximum bin number (user defined limit for the histogram plot), t_i is the time (or timesamp) of occurrence of i-th spike, $(t_i - t_{i-1})$ is the ISI between spikes i and i−1, N is the total number of spikes on the entire set of electrodes.

(iii) Inter Burst Interval (IBI): is the time interval between two consecutive bursts (spike sequences). It is computed for the whole set of electrodes burst data spike evaluated during the burst detection procedure. The IBI Histogram (IBIH) is computed evaluating the time (samples) intervals between every couple of consecutive bursts while time binning at user specified bin width:

$$IBIH\,(bin_k) = \sum_i \left[k - 1 \leq \frac{(t_{bi} - t_{b(i-1)})}{w} < k\right], \quad k = 1, \ldots, m; \; i = (2, B) \quad (3.3)$$

where w is the bin width, k is the k-th bin index, m is the maximum bin number (user defined limit for the histogram plot), t_{bi} is the time of occurrence of burst bi (first spike of the sequence), $(t_{bi} - t_{b(i-1)})$ is the IBI between bursts i and (i−1), B is the total number of bursts on the entire set of electrodes. Indeed, network's mean and standard deviation of the IBI is also provided.

(iv) Burst Duration (BD): is the sum of the ISIs of the sequence of consecutive spikes within the burst itself, i.e., the time interval between the first and the last spike in the spikes sequence. I applied an efficient strategy by storing only the first and last spike time stamps (and the number of spikes within the burst) allowing the computation of the BD as the difference between the two time stamps. The BD histogram (BDH) is constructed by summing the whole network's total number of bursts durations at a user specified time bin:

$$BDH(bin_k) = \sum_i \left[k - 1 \leq \frac{(t_{bi,n} - t_{bi_0})}{w} < k\right], \quad k = 1, \ldots, m; \; i = (2, B) \quad (3.4)$$

where w is the bin width, k is the k'th bin index, m is the maximum bin number (user set limit for the histogram plot), $t_{bi,0}$ is the time of occurrence of the first spike in burst bi (first spike of the sequence), $t_{bi,n}$ is the time of occurrence of the last spike in burst bi, B is the total number of bursts on the entire set of electrodes.

(v) Bursting statistics: SpiCoDyn provides the computation of single electrode's bursting rates (BR) and the overall network's mean bursting rate (MBR). It also provides the calculation of the percentage of random spikes (that can be alternatively expressed as the complementary of the percentage of bursts related spikes [56] for a single electrode and the entire network's mean and standard deviation values). Random spikes are the spikes that are not involved in the sequences within the bursts. Finally, the percentage of bursting time (with related network's mean value and standard deviation) is provided. It is defined as the fraction of the recording length (for each electrode) in which bursts occur:

$$\%\, coverage = L / \sum_i BD_i, \quad i = 1, \ldots, n \tag{3.5}$$

where L is the total recording's length, n is the total number of electrode's bursts, and BD_i is the i-th burst's duration.

3.6.9 *Comparison with Other Software*

I performed a detailed comparison between SpiCoDyn and SpyCode based on the times needed to perform the most resources-requiring dynamics analysis operations, and the computation of a subset of the main spiking/bursting parameters. Table 3.1 summarizes the results in terms of the time needed to complete each of the main steps of the reported set of operations, on a 10-min recording of a mature cortical neuronal network coupled to a 60-electrodes MEA. The software run under Windows 10 on 4 cores i7, 2.5 GHz, 16 GB of RAM and solid state disk. It is important to underline that while obtaining appreciably lower computational times, the results in terms of the produced output values of the procedures are identical compared to those obtained with SpyCode (some not significant differences emerged only due

Table 3.1 Comparison of the SpiCoDyn and SpyCode computation times evaluated for the processing of a neural network coupled to a 60 microelectrodes MEA (from MCS). All the computational tests have been performed on a 4 cores i7 2.5 GhZ, 16 GB of RAM and solid state disk. Reproduced from [4]

Algorithm	SpiCoDyn[s]	SpyCode [s]
Automatic threshold (30 windows of 200 ms)	1	10
Manual threshold (50 s interval)	2	10
Spike detection (SDDT)	2	15
Spike detection (PTSD)	12	25
Spike analysis (MFR + Raster Plot + ISI)	1 + 5 + 2	5 + 30 + 5
Burst analysis (MBR + MBD + IBI)	1 + 2 + 2	5 + 5 + 5

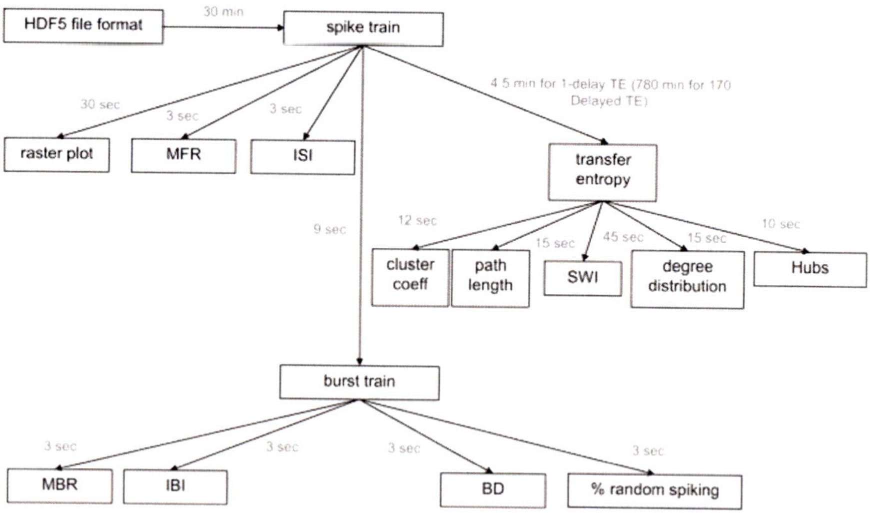

Fig. 3.25 Schematic representation and relative computational times of the algorithm used to analyze 30 min of spontaneous activity recorded by high density MEA with 4096 microelectrodes. All the computational tests have been performed on a 4 cores i7 2.5 GHz, 16 GB of RAM and solid state disk. Reproduced from [4]

to some customized strategies that I applied in the spiking and bursting statistics computation).

Spiking and connectivity analysis performed on HD-MEAs show fast computational time, considered the large amounts of calculations to be performed (Fig. 3.25). When considering spike detection (the most demanding computational task), SpiCoDyn requires about the same time that BrainWaveX software (3Brain) needs to run the same algorithm.

3.6.10 Use of SpiCoDyn to Process Long-Lasting Recordings on MEA-4k

In this section, I will report an example of SpiCoDyn's application to cortex neural networks coupled to the MEA-4k. As Fig. 3.25 shows, the analysis is composed of a part related to the dynamics and one related to the evaluation of the topological properties of the network. The total time required to analyze both the spiking and bursting dynamics, as well as the functional-effective topological properties is 812 min. It is worth noticing that the dynamical characterization requires less than 31 min: 30 min for the spike detection and 26 s to produce all the output (figures and text files). Most of the computational time is consumed by the functional-effective connectivity algorithm. However, it is worth to underline that it was necessary the computation of 170 TEs to cover a temporal window of 20 ms, while a single TE required only 4.5 min to be evaluated. Finally, the topological features extraction on the active electrodes (i.e. 1980 over 4096 electrodes) required only 95 s to be processed.

3.6.11 Dynamic Characterization

SpiCoDyn is independent from the acquisition system used to acquire the electrophysiological data; according to this, it uses the HDF5 standardized input data format. The 3Brain acquisition system allows to directly export the raw data in HDF5 format online during the recordings. Once the HDF5 files has been stored, the user can select the folder containing the raw data. In this way, SpiCoDyn permits to analyze in parallel all the experiments and phases contained in the input data folder. After reading the HDF5 files, the PTSD algorithm is used to spike detect the raw data. I report the results of a complete analysis for a neural network coupled to the MEA-4k. The mean MFR computed among all the active electrodes (1916 over 4096) was 0.55 $\pm$ 1.63 spikes/s. Figure 3.2a represents the ISIH, -where it can be noticed that most of the ISIs are smaller than 15 ms with an exponential-like decay. Then, I performed a bursting analysis using the algorithm described in *Burst Analysis*. Bursting features are characterized by means of the IBIH (Fig. 3.2b) and the mean burst duration histogram (Fig. 3.2c). As the table of the panel of Fig. 3.2d shows, bursting activity presents a mean BD of 139 $\pm$ 89 ms, a MBR of 1.74 $\pm$ 3.31 bursts/s and a percentage of random spiking of 51.26 $\pm$ 31.55%.

3.6.12 Connectivity Characterization

The connectivity characterization is reported in section *DTE application to experimental data.*

References

1. Fawcett T (2006) An introduction to ROC analysis. Pattern Recogn Lett 27:861–874
2. Ito S et al (2011) Extending transfer entropy improves identification of effective connectivity in a spiking cortical network model. PLoS ONE 6:e27431. https://doi.org/10.1371/journal.pone.0027431
3. Poli D, Pastore VP, Martinoia S, Massobrio P (2016) From functional to structural connectivity using partial correlation in neuronal assemblies. J Neural Eng 13:026023
4. Pastore VP, Godjoski A, Martinoia S, Massobrio P (2017) SPICODYN: a toolbox for the analysis of neuronal network dynamics and connectivity from multi-site spike signal recordings. Neuroinformatics 16:15–30. https://doi.org/10.1007/s12021-017-9343-z
5. van den Heuvel MP, Sporns O (2013) Network hubs in the human brain. Trends Cogn Sci 17:683–696. https://doi.org/10.1016/j.tics.2013.09.012

6. Downes JH et al (2012) Emergence of a small-world functional network in cultured neurons. PLoS Comput Biol 8:e1002522. https://doi.org/10.1371/journal.pcbi.1002522
7. Langer N, Pedroni A, Jäncke L (2013) The problem of thresholding in small-world network analysis. PLoS ONE 8:e53199. https://doi.org/10.1371/journal.pone.0053199
8. Bonifazi P et al (2009) GABAergic hub neurons orchestrate synchrony in developing hippocampal networks. Science 326:1419–1424
9. Pastore VP, Massobrio P, Godjoski A, Martinoia S (2018) Excitatory-inhibitory links and network topology in large scale neuronal assemblies from multi-electrode recordings. PLoS Comput Biol 14(8):e1006381. https://doi.org/10.1371/journal.pcbi.100638
10. Kreitzer AC (2009) Physiology and pharmacology of striatal neurons. Annu Rev Neurosci 32:127–147. https://doi.org/10.1146/annurev.neuro.051508.135422
11. Kaufman AM et al (2012) Opposing roles of synaptic and extrasynaptic NMDA receptor signaling in cocultured striatal and cortical neurons. J Neurosci 32:3992–4003. https://doi.org/10.1523/jneurosci.4129-11.2012
12. Marom S, Shahaf G (2002) Development, learning and memory in large random networks of cortical neurons: lessons beyond anatomy. Q Rev Biophys 35:63–87
13. Schröter M, Paulsen O, Bullmore ET (2017) Micro-connectomics: probing the organization of neuronal networks at the cellular scale. Nat Rev Neurosci 18:131. https://doi.org/10.1038/nrn.2016.182
14. Barabasi AL, Albert R (1999) Emergence of scaling in random networks. Science 286:509–512
15. Poli D, Pastore VP, Massobrio P (2015) Functional connectivity in *in vitro* neuronal assemblies. Front Neural Circ 9:57. https://doi.org/10.3389/fncir.2015.00057
16. P Massobrio L Arcangelis de V Pasquale HJ Jensen D Plenz 2015 Criticality as a signature of healthy neural systems Front Syst Neurosci 9 22https://doi.org/10.3389/fnsys.2015.00022
17. Schroeter MS, Charlesworth P, Kitzbichler MG, Paulsen O, Bullmore ET (2015) Emergence of rich-club topology and coordinated dynamics in development of hippocampal functional networks *in vitro*. J Neurosci 35:5459–5470. https://doi.org/10.1523/JNEUROSCI.4259-14.2015
18. Maccione A et al (2010) Experimental investigation on spontaneously active hippocampal cultures recorded by means of high-density MEAs: analysis of the spatial resolution effects. Front Neuroeng 3:4. https://doi.org/10.3389/fneng.2010.00004
19. Gerstein GL, Perkel DH (1972) Mutual temporal relationships among neuronal spike trains. Biophys J 12:453–473. https://doi.org/10.1016/S0006-3495(72)86097-1
20. Rosenberg JR, Amjad AM, Breeze P, Brillinger DR, Halliday DM (1989) The Fourier approach to the identification of functional coupling between neuronal spike trains. Prog Biophys Mol Biol 53:1–31
21. Brosch M, Schreiner CE (1999) Correlations between neural discharges are related to receptive field properties in cat primary auditory cortex. Eur J Neurosci 11:3517–3530. https://doi.org/10.1046/j.1460-9568.1999.00770.x
22. Eytan D, Minerbi A, Ziv N, Marom S (2004) Dopamine-induced dispersion of correlations between action potentials in networks of cortical neurons. J Neurophysiol 92:1817–1824
23. Quian Quiroga R, Kreuz T, Grassberger P (2002) Event synchronization: a simple and fast method to measure synchronicity and time delay patterns. Phys Rev 66:041904
24. Bedenbaugh P, Gerstein GL (1997) Multiunit normalized cross correlation differs from the average single-unit normalized correlation. Neural Comput 9:1265–1275
25. Salinas E, Sejnowski TJ (2001) Correlated neuronal activity and the flow of neural information. Nat Rev Neurosci 2:539–550
26. Aertsen AMHJ, Gerstein GL (1985) Evaluation of neuronal connectivity: sensitivity of cross-correlation. Brain Res 340:341–354. https://doi.org/10.1016/0006-8993(85)90931-X
27. Melssen WJ, Epping WJ (1987) Detection and estimation of neural connectivity based on crosscorrelation analysis. Biol Cybern 57:403–414
28. Dunn B, Mørreaunet M, Roudi Y (2015) Correlations and functional connections in a population of grid cells. PLoS Comput Biol 11:e1004052. https://doi.org/10.1371/journal.pcbi.1004052

29. Aertsen A et al (1991) Neural interactions in the frontal cortex of a behaving monkey: signs of dependence on stimulus context and behavioral state. J Hirnforsch 32:735–743
30. Schneidman E, Berry MJ, Segev R, Bialek W (2006) Weak pairwise correlations imply strongly correlated network states in a neural population. Nature 440:1007–1012. http://www.nature.com/nature/journal/v440/n7087/suppinfo/nature04701_S1.html
31. Cutts CS, Eglen SJ (2014) Detecting pairwise correlations in spike trains: an objective comparison of methods and application to the study of retinal waves. J Neurosci 34:14288–14303. https://doi.org/10.1523/jneurosci.2767-14.2014
32. Salinas E, Sejnowski TJ (2000) Impact of correlated synaptic input on output firing rate and variability in simple neuronal models. J Neurosci 20:6193
33. Garofalo M, Nieus T, Massobrio P, Martinoia S (2009) Evaluation of the performance of information theory-based methods and cross-correlation to estimate the functional connectivity in cortical networks. PLoS ONE 4:e6482. https://doi.org/10.1371/journal.pone.0006482
34. Eichler M, Dahlhaus R, Sandkuhler J (2003) Partial correlation analysis for the identification of synaptic connections. Biol Cybern 89:289–302
35. Newman ME (2006) Modularity and community structure in networks. Proc Natl Acad Sci USA 103:8577–8582
36. Kanagasabapathi TT et al (2012) Functional connectivity and dynamics of cortical-thalamic networks co-cultured in a dual compartment device. J Neural Eng 9:036010. https://doi.org/10.1088/1741-2560/9/3/036010
37. Berdondini L et al (2009) Active pixel sensor array for high spatio-temporal resolution electrophysiological recordings from single cell to large scale neuronal networks. Lab Chip 9:2644–2651
38. Buchs P-A, Muller D (1996) Induction of long-term potentiation is associated with major ultrastructural changes of activated synapses. Proc Natl Acad Sci USA 93:8040–8045
39. Eversmann B et al (2003) A 128 x 128 CMOS biosensor array for extracellular recording of neural activity. IEEE J Solid-State Circ 38:2306–2317
40. Muller J et al (2015) High-resolution CMOS MEA platform to study neurons at subcellular, cellular, and network levels. Lab Chip 15:2767–2780. https://doi.org/10.1039/c5lc00133a
41. Viswam V, Dragas J, Muller J, Hierlemann A In: IEEE international solid-state circuits conference. IEEE, pp 394–396
42. Pastore VP, Poli D, Godjoski A, Martinoia S, Massobrio P (2016) ToolConnect: a functional connectivity toolbox for *in vitro* networks. Front Neuroinformatics 10:13. https://doi.org/10.3389/fninf.2016.00013
43. Overbey LA, Todd MD (2009) Dynamic system change detection using a modification of the transfer entropy. J Sound Vib 322:438–453. https://doi.org/10.1016/j.jsv.2008.11.025
44. HDFGroup (2013) Hierarchical data format
45. Bennett K, Robertson J (2011) In: Ionescu CM (ed) MATLAB—a ubiquitous tool for the practical engineer
46. Bologna LL et al (2010) Investigating neuronal activity by SPYCODE multi-channel data analyzer. Neural Netw 23:685–697. https://doi.org/10.1016/j.neunet.2010.05.002
47. Vato A et al (2004) Spike manager: a new tool for spontaneous and evoked neuronal networks activity characterization. Neurocomputing 58:1153–1161
48. Maccione A et al (2009) A novel algorithm for precise identification of spikes in extracellularly recorded neuronal signals. J Neurosci Methods 177:241–249
49. Tam DC (2002) An alternate burst analysis for detecting intra-burst firings based on inter-burst periods. Neurocomputing 44–46:1155–1159
50. Kapucu FE et al (2012) Burst analysis tool for developing neuronal networks exhibiting highly varying action potential dynamics. Front Comput Neurosci 6:38. https://doi.org/10.3389/fncom.2012.00038
51. Mazzoni A et al (2007) On the dynamics of the spontaneous activity in neuronal networks. PLoS ONE 2:e439. https://doi.org/10.1371/journal.pone.0000439
52. Pasquale V, Martinoia S, Chiappalone M (2009) A self-adapting approach for the detection of bursts and network bursts in neuronal cultures. J Comput Neurosci 29:213–229. https://doi.org/10.1007/s10827-009-0175-1

53. Chiappalone M et al (2005) Burst detection algorithms for the analysis of spatio-temporal patterns in cortical networks of neurons. Neurocomputing 65–66:653–662
54. Bal-Price AK et al (2010) In vitro developmental neurotoxicity (DNT) testing: relevant models and endpoints. NeuroToxicology 31:545–554. https://doi.org/10.1016/j.neuro.2009.11.006
55. Hogberg HT et al (2011) Application of micro-electrode arrays (MEAs) as an emerging technology for developmental neurotoxicity: evaluation of domoic acid-induced effects in primary cultures of rat cortical neurons. NeuroToxicology 32:158–168. https://doi.org/10.1016/j.neuro.2010.10.007
56. Grace AA, Bunney BS (1984) The control of firing pattern in nigral dopamine neurons: burst firing. J Neurosci 4:2877

Conclusion

It is fundamental to elaborate research strategies aimed to a comprehensive structural description of neuronal interconnections as well as the networks' elements forming the human connectome [1]. The connectome will significantly increase our understanding of how functional brain states emerge from their underlying structural substrate, and will provide new mechanistic insights into how brain function is affected if this structural substrate is disrupted. The final goal of a connectivity analysis is the reconstruction of the human connectome, thus, the application of statistical measures to the in vivo model in both physiological and pathological states. Since the system under study (i.e. brain areas, cell assemblies) is highly complex, to achieve the purpose described above, it is useful to adopt a reductionist approach. During my Ph.D. work, I focused on a reduced and simplified model, represented by neural networks chronically coupled to MEAs. I developed and optimized statistical methods to infer functional connectivity from spike train data. In particular, I worked on correlation-based methods: cross-correlation and partial correlation, and information-theory based methods: Transfer Entropy (TE) and Joint Entropy (JE). More in detail, my aim was to reconstruct the functional-effective connectivity, trying to establish a relationship between dynamics and connectivity, in order to understand if and in which way these parameters influence the topological (functional) organization of a neural network. This information could be very useful to help in understanding the mechanisms underlying the superior cognitive functions (e.g., learning, memory, movement) in both physiological and pathological states. In order to topologically characterize neural networks, it is important to use high resolution acquisition system, having few neurons (1–3) per electrode. Thus, my PhD's aim has been applying functional connectivity methods to neural networks coupled to MEA-4K (3Brain APS with 4096 electrodes). To reliably reconstruct the functional topological organization of neural networks it is fundamental to detect and analyze both excitatory and inhibitory links. In fact, inhibitory links not only represent a consistent part of the total functional links (~25% [2]), but they can be fundamental in orchestrating the synchrony and shaping the connectivity [3]. No published work

V. P. Pastore, *Estimating Functional Connectivity and Topology in Large-Scale Neuronal Assemblies*, Springer Theses, https://doi.org/10.1007/978-3-030-59042-0

about functional-effective connectivity estimation on multiple neuronal spike trains provides a well-defined computational method to identify inhibitory connections. Starting from the standard definition of the cross-correlation [4] (cf., Methods), I adopted the normalization approach described in [5, 6], to obtain the "raw" Normalized Cross-Correlation Histogram (NCCH). I formalized my hypothesis that, the extraction of negative peaks (rather than troughs) obtained through a simple filtering operation on the NCCH and followed by distinct thresholding operations for excitatory and inhibitory connections, permits to identify a significant percentage of inhibitory connections with a high level of accuracy. In fact, theoretically, cross-correlation is able to detect both an increase and a decrease of the synchrony between the spike trains relative to putative interconnected neurons. However, in real experimental data, the cross-correlogram is very jagged making difficult the detection of small peaks and troughs, and, apart specific conditions (i.e., high and tonic firing rate) [7] hindering the detection of inhibition. My approach has consisted in a simple post processing of the cross-correlation histogram, obtaining what I called Filtered and Normalized Cross-Correlation Histogram. The FNCCH was able to detect with very high precision both the excitatory and the inhibitory links in in silico neural networks. I applied the FNCCH algorithm to neural networks coupled to the MEA-4k, trying to extract the topological organization. I found that for cortical networks coupled to MEA-4k devices it is possible to see the emergence of a clear small-world (SW) topology. With the measurements performed with MEA-4k's I can state that, both inhibitory and excitatory links with their small world index, SWI >> 1 (9.2 ± 3.5 for the inhibitory links and 5.2 ± 2 for the excitatory ones) contribute to 'segregation' with the emergence of small world networks. Moreover, inhibitory links, with their somehow longer connections, might contribute also to network 'integration' (i.e., communication among the SWs). Such a result confirms the importance to perform functional connectivity analysis using a method able to detect both excitatory and inhibitory links, above all when the final aim is to characterize the topology of a neural network. When analyzing similar cortical networks but coupled to the MEA-60 devices, no SW topology is identified; such networks seem to be characterized by a sub-random topology with a SWI of 0.4 ± 0.1 for the excitatory and 0.2 ± 0.2 for the inhibitory links. These cortical networks are of the same type as the ones coupled to the MEA-4k (i.e., similar density of neurons, same age, same culture medium) and the apparent estimated random topology should be attributed to the low number of recording sites (i.e., 60 channels) that are not enough to reliably infer topological features (see section *FNCCH results*). Such a result confirms the importance to consider neural networks coupled to high resolution acquisition systems (MEA-4k). Both the results should be confirmed by applying the FNCCH algorithm to process more neural networks not only in physiological state, but also in a pathological state that could alter the topological organization. However, the use of MEA-4k produces a huge amount of data, requiring the development of ad-hoc computational efficient software platforms to process and analyze such data [8]. For this reason, I developed the software package SPICODYN, aiming to be a possible solution of the aforementioned big data problem, by introducing new software tools, independent from the acquisition systems, able to process in reasonable computational times, discrete time series coming from multi-site spike signal recordings. The philosophy under this project was to develop and share with the scientific community an open source platform that allows performing spike analysis by including new tools, to enhance the repertoire of algorithms. The current version of SpiCoDyn embeds a collection of functions (correlation and information-theory based) to estimate functional-effective connectivity, part of which directly come from the software ToolConnect [9], (https://www.nitrc.org/projects/toolconnect/). The connectivity analysis section has been revised and expanded introducing the implementation of a customized fully optimized version of the TE algorithm, tailored to be used on time stamps data extracted from multi-unit spike train recordings. In addition, SpiCoDyn implements an extension of the TE method to deal with multiple time delays (temporal extension) and with multiple binary pattern (high order extension), the FNCCH and the JE. Another feature of the developed software is the possibility to characterize the spiking and bursting dynamics, starting from raw data coming from different acquisition systems. By exploiting the fact that several commercial acquisition systems (e.g., 3Brain, Multi-Channel Systems) use the HDF5 file format for storing and managing data, I realized a tool that, starting from such metafiles, performs spike detection and then analysis of spiking and bursting activity. Nonetheless, also a specific analysis of the topological properties (i.e., connectivity properties) can be performed by exploiting the functional connectivity maps derived from the available algorithms for inferring functional-effective connectivity (e.g., cross-correlation, partial-correlation, joint entropy, transfer entropy). The current release of the software presents also two main limitations: the first one is the lack of algorithms to process the stimulus-evoked activity (e.g., Post-Stimulus Time Histogram, PSTH). The second one is that all the algorithms work only on spike trains data (i.e., point process). On the other hand, it is worth underlining that most of the fundamental optimization strategies applied in the implementation of the connectivity algorithms are based on the use of spike trains data, and therefore applicable only because of working with this kind of input data. Finally, it is important to underline that SpiCoDyn is available to the scientific community and it has been programmed to be adapted, modified and extended by the interested researchers.

Appendix

Computational Model

I used computational model to assess the connectivity methods accuracies (cf., Fig. A1). The network model is made up of 1000 neurons randomly connected. The dynamics of each neuron is described by the Izhikevich equations [10]. In the actual model, two families of neurons have been taken into account: regular spiking and fast spiking neurons for excitatory and inhibitory populations, respectively. The ratio of excitation and inhibition has been set to 4:1 as experimentally founded in cortical cultures [2]. In the model, each excitatory neuron receives 100 connections from excitatory and/or inhibitory neurons, while each inhibitory neuron receives 100 input only from excitatory neurons). Autapses are not allowed. All the inhibitory connections introduce a delay equal to 1 ms, while excitatory ones range from 1 to 20 ms. Synaptic weights are extracted from a Gaussian distribution with mean equal to 6 and -5 for excitatory and inhibitory weights. Standard deviations have been set to 1. Excitatory weights evolve following the spike timing dependent plasticity (STDP) rule with a time constant equal to 20 ms [11]. The spontaneous activity of the neuronal network has been generated by stimulating a randomly chosen neuron at each time stamp injecting a current pulse extracted from a normal distribution ($I_{stm,exc} = 11 \pm 2$; $I_{stm,inh} = 7 \pm 2$). The network model has been implemented in Matlab (The Mathworks, Natick, MA, USA), and each run simulates 1 h of spontaneous activity.

V. P. Pastore, *Estimating Functional Connectivity and Topology in Large-Scale Neuronal Assemblies*, Springer Theses, https://doi.org/10.1007/978-3-030-59042-0

V. P. Pastore, *Estimating Functional Connectivity and Topology in Large-Scale Neuronal Assemblies*, Springer Theses, https://doi.org/10.1007/978-3-030-59042-0

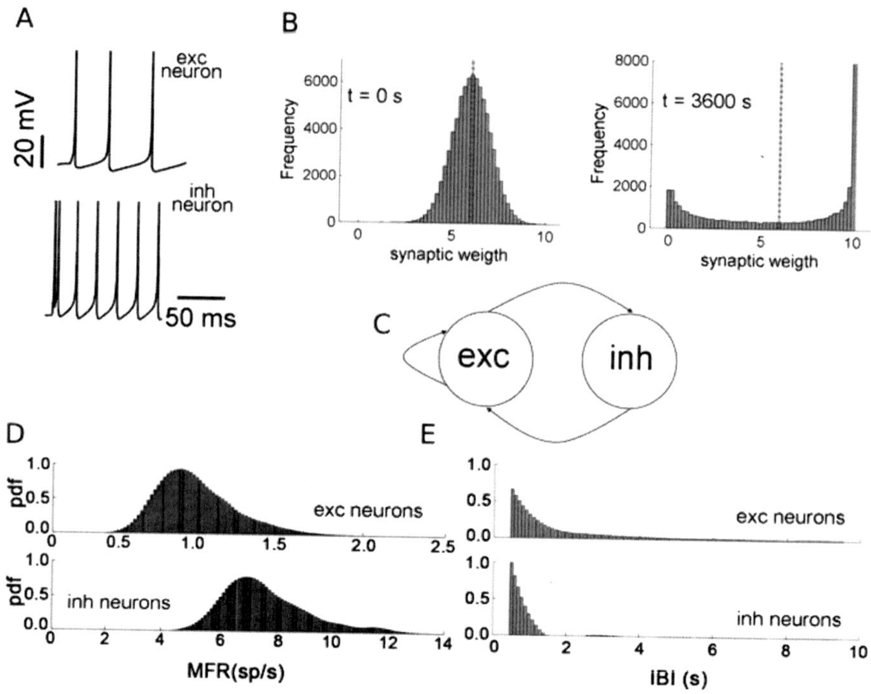

Fig. A1 Computational model features and simulation results. **a** electrophysiological patterns of excitatory (top) and inhibitory (bottom) neurons. **b** excitatory synaptic weights distribution at $t = 0$ (left side) and at the end of the simulation (right side). **c** sketch of the permitted connections among the excitatory and inhibitory populations. **d** MFR distributions. **e** IBI distributions. Adapted from [12]

References

1. Sporns O, Tononi G (2002) Classes of network connetivity and dynamics. Complexity 7:28–38
2. Marom S, Shahaf G (2002) Development, learning and memory in large random networks of cortical neurons: lessons beyond anatomy. Q Rev Biophys 35:63–87
3. Bonifazi P et al (2009) GABAergic hub neurons orchestrate synchrony in developing hippocampal networks. Science 326:1419–1424
4. Salinas E, Sejnowski TJ (2001) Correlated neuronal activity and the flow of neural information. Nat Rev Neurosc 2:539–550
5. Brosch M, Schreiner CE (1999) Correlations between neural discharges are related to receptive field properties in cat primary auditory cortex. Eur J Neurosci 11:3517–3530. https://doi.org/10.1046/j.1460-9568.1999.00770.x
6. Eytan D, Minerbi A, Ziv N, Marom S (2004) Dopamine-induced dispersion of correlations between action potentials in networks of cortical neurons. J Neurophysiol 92:1817–1824
7. Aertsen AMHJ, Gerstein GL (1985) Evaluation of neuronal connectivity: sensitivity of cross-correlation. Brain Res 340:341–354. https://doi.org/10.1016/0006-8993(85)90931-X
8. Buzsaki G (2004) Large-scale recording of neuronal ensembles. Nat Neurosci 7:446–451
9. Pastore VP, Poli D, Godjoski A, Martinoia S, Massobrio P (2016) ToolConnect: a functional connectivity toolbox for *in vitro* networks. Front Neuroinform 10. https://doi.org/10.3389/fninf.2016.00013
10. Izhikevich EM (2003) Simple model of spiking neurons. IEEE Trans Neural Netw 14:1569–1572
11. Song S, Miller KD, Abbott LF (2000) Competitive Hebbian learning through spike-timing-dependent synaptic plasticity. Nat Neurosci 3:919–926
12. Pastore VP, Massobrio P, Godjoski A, Martinoia S (2018) Excitatory-inhibitory links and network topology in large scale neuronal assemblies from multi-electrode recordings. PLoS Comput Biol 14(8):e1006381. https://doi.org/10.1371/journal.pcbi.100638

V. P. Pastore, *Estimating Functional Connectivity and Topology in Large-Scale Neuronal Assemblies*, Springer Theses, https://doi.org/10.1007/978-3-030-59042-0